CROSS-TRAINING FOR FIRST RESPONDERS

CROSS-TRAINING FOR FIRST RESPONDERS

GREGORY S. BENNETT

CRC Press
Taylor & Francis Group
Boca Raton   London   New York

CRC Press is an imprint of the
Taylor & Francis Group, an **informa** business

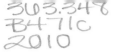

CRC Press
Taylor & Francis Group
6000 Broken Sound Parkway NW, Suite 300
Boca Raton, FL 33487-2742

© 2010 by Taylor and Francis Group, LLC
CRC Press is an imprint of Taylor & Francis Group, an Informa business

No claim to original U.S. Government works

Printed in the United States of America on acid-free paper
10 9 8 7 6 5 4 3 2 1

International Standard Book Number: 978-1-4398-2653-9 (Hardback)

Library of Congress Cataloging-in-Publication Data

Bennett, Gregory.
   Cross-training for first responders / Gregory Bennett.
      p. cm.
   Includes bibliographical references and index.
   ISBN 978-1-4398-2653-9 (hardcover : alk. paper) 1. Emergency management--United States. 2. First responders--Training of--United States. I. Title.

HV551.3.B46 2010
363.34'880683--dc22
                                                           2010010971

Visit the Taylor & Francis Web site at
http://www.taylorandfrancis.com

and the CRC Press Web site at
http://www.crcpress.com

Dedication

This book is dedicated to my beautiful and loving wife, Mary Pat. During the many years of our lives together, she has tolerated my being away from home for assorted reasons pertaining to my job and my volunteer endeavors. She has always been understanding and supportive of all my dreams and aspirations. She is my best friend and the love of my life. Without her caring concern and support, many of my achievements would not have come to fruition. Also, I dedicate this book to my parents and my grandfather, who instilled in me the morals, values, and work ethic that have allowed my life to follow its current course. Finally, I thank my stepson, John, and all of my colleagues and peers, who have always been supportive of my various projects and offered constant encouragement, advice, support, and assistance.

Contents

Preface

The United States suffered a tremendous loss of lives in four separate and almost simultaneous attacks on September 11, 2001. The world watched in horror as the World Trade Center buildings collapsed after being struck by commercial airplanes; they also saw the Pentagon burning and the remains of another commercial aircraft that was forced to crash after the passengers tried to take control from hijackers. In the hours and days following these cowardly attacks, professional responders and ordinary citizens raced to the various locations trying to help the victims however they could. While the emergency service community was able to bring training, tools, and equipment to the incident locations, they were also attempting techniques they were never trained for, and were working without the correct equipment.

In the years since these horrific incidents, significant progress has been made in many of the areas deemed deficient after the initial response in 2001. Reviewing authorities cited many areas where the intelligence and response communities were caught unprepared. Some of these issues included information sharing among the various intelligence gathering agencies, and communications issues among firefighters, law enforcement, and many other responders. Money was allocated to address these areas, and now information flows more fluidly among the entities that generate and review the collected materials. The government awarded contracts to different companies to develop technology to bridge the communications gaps discovered during and after the attacks. While the response community will certainly embrace new technologies, there are other areas where the responders could have benefited. Further, the many other sources of potential "first responders" has not been fully explored, developed, or trained.

Traditionally, each response community has focused nearly exclusively on techniques required for its unique discipline. Firefighters focused on their traditional response areas of suppression, search and rescue, and similar issues. Similarly, the emergency medical responders' training focused on triage, treatment, patient packaging, and transport. One of the lessons learned from the September 11th response is that in times of crisis, incident managers may have untrained, but willing "first responders." What this means is that people who are not normally considered to be first responders may actually be on-scene and attempting to assist victims well before professional responders arrive. These potential responders include security guards, school teachers and administrators, airport personnel, school bus drivers, and others. Depending on where an incident occurs, some of these people may already be present and begin to take steps to mitigate the negative consequences. These potential responders will be attempting operations for which they may have no training, experience, or equipment. This set of circumstances is similar to the actions taken by some of the professional responders on September 11th. Recognizing that there are certain civilian populations who may become reluctant first responders, there needs to be a training program that can be presented and reviewed by these civilians. The training that each of these other miscellaneous or non-traditional responders should be given will be discussed at length in this book. We cannot ignore the actions that may be attempted by these populations; they should receive some form of training.

The same argument can be made of our professional responders. There is great potential for response communities to venture across their traditional boundaries or duties in the event of another terrorist attack. Law enforcement personnel will be the responders in need of the most cross-training for a number of reasons. They will most likely be the first arriving professional responders, and there will be an inclination to attempt techniques, even without prior training. An obvious example involves police officers racing into burning buildings without specific training regarding what they may experience inside, or training on how to locate and remove victims. In the new age of terrorism, all emergency responders will be expected to cross into the traditional workplace held by other responders. Financial and manpower crises across the country also make a strong case for the professional responders to be prepared to do more. In fact, almost everywhere, there are more tasks to be performed, and fewer, adequately-trained responders assigned to complete them. If society demands extra efforts from the response communities, it should also demand that the responders be trained in the new areas they will be thrust into. This training can also be offered to political leaders, Office of Emergency Management (OEM) staff, public health professionals, and hazardous materials teams. Each of the groups mentioned could benefit greatly from training in areas where they would normally ask for assistance, rather than mitigate the emergency themselves. It is also important for our political leaders to see exactly what the issues faced by the responders are, in order for them to understand funding and staffing requests made by response administrators.

Both volunteer *and* career professionals respond to calls for service every day. Both volunteer and career personnel do a very commendable job of answering these calls for service. However, there is a certain amount of tension between the ranks of these two groups. Once they are operating together, there is an uneasy truce, but we need to find a way for all of our responders to simply get along. Specific examples of when, where, and why there are tensions between the volunteer and career responders will be identified and discussed in this book. This situation is generally controlled, but there is a potential for an explosion between the responder groups. This needs to be curtailed.

Once a need for training is recognized, several obstacles need to be dealt with prior to implementing the training. The specific materials or courses to be presented need to be agreed upon, and either developed or located for delivery to those in need. Should there be no existing training curricula in place, a course or class will need to be developed. Trainers or instructors need to be located, and the training goals need to be identified and reviewed by the trainer. Certain time and cost requirements will need to be addressed as well. Career agencies must find a way to reduce overtime and related costs in order to be cost effective when presenting new training materials. Volunteer agencies will face the challenge of getting all of their personnel together to be trained. Also, if agency policy does not make the new training mandatory, compliance may become an issue for responders failing to take the prescribed class. Training options, including federal trainers who will come to your location, will be reviewed and sources for these resources will be listed also. These contractors may be able to provide the training required at no cost to the agency in need of training. These types of innovative training strategies will be reviewed.

Training in these new response areas may result in a requirement for new, specialized equipment along with classroom instruction. With shrinking budgets, innovative ideas to acquire the new training and equipment need to be developed. There are certain grant opportunities for the emergency response community. These grants, as well as application tips, will be explored in the following chapters, and specific examples of narrative do's and don'ts will be reviewed. Grant opportunities are very specific, and close attention must be paid to all of the instructions in order to be considered a successful grant recipient.

Another option for training your personnel is using virtual academies, or online courses. There are pros and cons to these virtual academy classes, and these will be discussed at length. Caution must be exercised to ensure that the personnel taking the classes do so without assistance from other personnel. Many of the online classes require tests for completion credits, and others do not. A big drawback in these classes is that there is no forum for the students to ask questions in cases where the material is confusing, or not grasped by the students. The other online trend for emergency service personnel is distance-learning offerings by many colleges and universities. As with virtual academies, there are pros and cons associated with these classes. This phenomenon will be reviewed in this volume and certain tips offered.

Should we convince our leaders to provide cross-training for our responders, there needs to be a mechanism to test the abilities of the responders to apply these new techniques. There are two generally accepted techniques to accomplish this task. The first is called a "tabletop" drill. This exercise involves fewer personnel than a traditional hands-on drill. There are tremendous amounts of planning required in advance. This type of drill allows the supervisors of the agency conducting the exercise to be tested on real-world events or incidents in a more pristine environment.

There is an entire chapter in the book that will address in great detail the steps involved in planning and conducting the tabletop exercise.

Logic would hold that after a tabletop exercise has been completed, a more thorough drill be conducted to test the decisions that were made during the tabletop scenario. The practical application of the lessons learned from the tabletop can be beneficial, or the planners can choose a different scenario and rotate key participants into new roles. This entire process will be explored and reviewed. Obviously, more participants are required for the practical drill, and, therefore, there will be greater associated costs.

The problems associated with a lack of cross-training will be discussed. Options for overcoming these issues will be presented. A problem that is unrecognized can have serious implications during an actual emergency. A problem that is identified and known presents different liabilities if it is ignored. This book will present arguments for training many different traditional and non-traditional first responders. Training classes and delivery options will be presented and discussed in detail and at length. The emergency response community will be crossing into new areas of service, and it will need to be prepared to act. Preparation begins with identification of the problem and then taking steps to correct it. Training is critical in resolving the problems faced by first responders.

The problem is now fully identifiable and solutions are available. Failure to act on these issues may result in a poor response to another major incident. This scenario will not be tolerated by the public who have witnessed response failures too often in the past.

About the Author

Gregory Bennett has been involved in emergency service since 1986. He is a career law enforcement officer, and has risen to the rank of Lieutenant for the Middlesex County (New Jersey) Sheriffs Department, where he has been employed since 1988. Among his duties, Lt. Bennett conducts all training related to homeland security for his agency. He is also the commander of the Sheriff's forty-member component of the New Jersey Urban Area Security Initiative Region Law Enforcement Rapid Development Team. He is certified as a New Jersey firearms instructor, and has instructed others on the use of force and vehicle pursuit for his agency, in addition to many other law enforcement topics. He also has work experience as a corrections officer and a juvenile and adult parole officer in New Jersey. He has lectured on the subject of cross-training first responders in Atlantic City, New Jersey and Washington, DC.

Additionally, Greg has more than twenty-four years of experience as a volunteer firefighter. He served as a Chief of the Helmetta (New Jersey) Fire Department from 2005–2007. He has served as a chief level officer for more than ten years. He also works as a part-time instructor at the Middlesex County Fire Academy, having served there since 2003. At the academy, he is assigned to fire-recruit training as well as in-service training and drills for veteran firefighters. Between 1999 and 2005, Greg was a per-diem firefighter and fire lieutenant for the Monroe Township Fire District #3 in Monroe, New Jersey. He has extensive experience in incident command and is certified to teach through the ICS-400 level, as well as I-700 NIMS. Finally, Greg has served as a deputy coordinator in the Office of Emergency (OEM) in Helmetta since 2007.

Greg is certified as New Jersey Firefighter 2, Fire Officer 1, New Jersey Hazardous Materials-Operations Level, Incident Management Level 3, and as a Level 2 Fire Service Instructor certified by the state of New Jersey. He is certified to instruct many classes for both fire and law enforcement personnel.

Greg holds a BS in criminal justice from the University of Dayton, Ohio, earned in 1986. He is married to Mary Patricia and has a stepson John, who is also a volunteer fire officer.

Chapter 1

Basic Concepts in Cross-Training

In the post 9/11 era, the United States has pumped billions and billions of dollars into a general field we will refer to as homeland security. Prior to that fateful date in history, we, as a country, were generally reactive responders in many areas relating to emergency response. Since then, there has been a significant shift in our response ideology. This new philosophy has us moving away from reactivity and speeding toward proactivity. The events and aftermath of 9/11 shed light on many issues that were evident and rampant throughout our emergency service communities. Civic leaders on all levels realized that a change in the response ideology was needed to better protect our citizens. While the attempt to accomplish this reversal of policy has been commendable, we have "missed the boat" in many areas.

Since this shift toward operational proactivity, we have seen the creation and development of new training programs: the introduction of a new cabinet-level federal department, online training courses, and think tanks that pop up like dandelions on a poorly maintained yard. As usual, the knee jerk reaction of government leaders was to respond to the first significant international terrorist event on our soil by throwing great amounts of money at the problem. Our political leaders seem to be of the opinion that if the population sees federal funds applied to a problem, then the problem will be quickly and completely eradicated. Nothing could be further from the truth. The speed at which money was appropriated and disbursed after 9/11 was astounding. Superficially, it would seem that not enough time was allotted to analyze the problems and how best to overcome them. Funds were made available by the federal government, the Department of Homeland Security was created, and states, counties and local governing bodies feasted on the sudden availability of federal dollars.

A more prudent approach should have been explored and implemented prior to simply making the federal dollars available for certain projects and equipment for our first responders. Programs such as National Incident Management System (NIMS), or the National Interagency Incident Management System (NIIMS) were quickly developed and distributed for the training of first responders. Federal funding opportunities became quickly linked to ensuring compliance with the NIMS/NIIMS doctrines. Failure to comply with these doctrines eliminated your agency from qualifying for the sudden influx of federal funds for homeland security. Sometimes, the application of these funds was not properly overseen. In Newark, New Jersey, for example, Mayor Sharpe James purchased new air conditioned garbage trucks with homeland security funds allocated to his city.[1] His reasoning was that if there were another terrorist attack on the scale of 9/11, there would be a need to clean up the garbage. Clearly, this was a severe misappropriation of federal funds earmarked for hardening or preparing his city for a potential terrorist attack.

Responding to Tragedy

After the 9/11 attacks in New York City, emergency responders from all over America rushed to support the responders and victims at the World Trade Center. Many of the police, fire, and Emergency Medical Services (EMS) crews, which responded to the area known as ground zero, were self-dispatched. The need to help overcame the training systems of mutual aid and incident command with which many of these responders were familiar. The intentions were pure and good, but the results were chaotic, to say the least.

When responders self-dispatch, there is no accountability system in place at the command post, which is the physical location of those charged with managing the incident. As an incident commander myself, I can attest to this being a totally unacceptable situation. Incident managers need to have strict and accurate accounting of all those operating at any incident. Not only were there self-dispatched resources flooding lower Manhattan, the resources themselves were not necessarily the type needed by the people managing the incident and overseeing the rescue efforts. Obviously, after the catastrophic collapse of two 110-story office buildings, responders with a specific skill set were required at the scene. These resources are known as Urban Search and Rescue (USAR) Teams. The USAR Teams were requested early on, and these resources were immediately assembling in their respective home states readying for deployment. These federally certified teams have special training related to, among other things, collapse rescue, hazardous materials, rescue and communications, and they bring specialty equipment with them. Generally untrained and unequipped responders from across the emergency services community rushing to the scene were more likely to add to the confusion at the incident than they were to provide necessary assistance to those in charge of incident management.

What Can Our Responders Realistically Do?

As lay people, the citizens of this country are mostly unaware of the capabilities, limitations, training discrepancies, lack of standardized training and accreditation or certification, and terminology differences faced within the response communities. I have observed these discrepancies first-hand and often wondered why this was the case. NIMS has tried to address these issues, but sadly, it will be years before any nation-wide changes are recognized and implemented. The NIMS program has many different components and makes many suggestions in the areas of training, standards, credentialing, and certifying both responders and equipment. Currently, there is no set federal standard that covers all of the areas addressed by NIMS. The emergency response community uses many of the same tools and equipment, but they do not share the same terminology all across America. This can cause confusion and may actually hinder mitigation efforts if responders from one region are assisting in another. If the responders need a piece of equipment, but it has a different name in their home area, the proper resource may already be present on-scene, or it may never get there.

Terminology Differences Are Glaring Issues

In the world of firefighting, there is a difference in terminology for resources on the east and west coasts. To a child, or to many adults, the terms fire engine and fire truck are synonyms, and, therefore, interchangeable. However, nothing could be further from the truth. A fire engine and a fire truck are indeed two separate pieces of firefighting equipment, with virtually no common ground. Fire engines have a pump and carry hoses and water. They are equipped with fire pumps of varying capacities to pump the water through different-sized hoses to extinguish a fire. A fire truck conversely carries no water, hoses, or pumps. A truck has an aerial device, whether it is a ladder or a platform or "bucket." The function of the ladder truck is to provide the crew with the ability to reach heights they cannot reach with a ground or extension ladder. Trucks or aerials are used for the rescue of trapped victims and as part of a *defensive* fire attack, which means that there are no firefighters working inside the structure. Is this difference important? Speaking as a firefighter and an incident commander, I say it most certainly is. If there is a tactical need for rescuers to get onto the roof of an apartment building with three stories, they would need a truck, not an engine. Similarly, if there is a car on fire, there would be a need for water, hose, and pump on the engine to suppress the flames. Certainly, to an untrained citizen or layman, this is a subtle difference in terminology, but it is a significant one to a fire officer ordering specific resources to respond to an incident.

Also, consider terminology differences within the fire discipline faced by its firefighters. On the east coast, for example, a tanker is a vehicle driven on surface roads, carries a minimum of 1,500 gallons of water, and is capable of pumping water at a minimum flow rate of 750 gallons per minute. On the west coast, a tanker is a fixed-wing aircraft (an airplane) that is capable of delivering water, typically in wildfire applications. Therefore, a fire officer "helping" in California when he normally works in New Jersey can be asking for the wrong resource for the tactical need at the incident. This would be understandable for a person with no training, but it is completely unacceptable for a trained firefighter to make this mistake (see Figures 1.4 and 1.5).

Figure 1.1 We will all recognize this as a school bus, used for transporting children to and from school and school activities. "Bus" is sometimes used by first responders to request a vehicle to transport injured and sick people. (Image copyright Regein Paasser 2009, under license from Shutterstock.com.)

As we can see, if a firefighter can be uncertain about what to call a piece of fire-fighting equipment, what would we expect for a policeman or an emergency medical technician (EMT) to ask for under the same circumstances? There are other examples for you to consider. After a motor vehicle collision where the cars are unable to be driven away, what resource would you call for? Do you need a flatbed truck, a tow truck, a hook, or a wrecker? These are all basically the same resource. All emergency responders know what resource they need to accomplish vehicle removal, but the resource has many names. Similarly, at the same incident, if there are injuries, what resource is requested? Do we need a bus, ambulance, Basic Life Support (BLS) Unit, Advanced Life Support (ALS) Unit, or a rig (Figure 1.1 and Figure 1.2)? Again, these are all basically the same thing—a vehicle used to transport someone injured or in need of definitive medical care. I have referenced a few examples where resources are the same, but have different names. Terminology is but one small area in the responders' tool kits. If this basic subject can cause confusion, what will become of them in the face of complex issues? If they are unfamiliar with a scenario, will they be able to render safe and efficient aid and request the correct resource using the proper terminology? Are they aware of what tools and equipment other responders can bring to the incident to ensure a prompt, professional, and efficient resolution?

Helping Our Own

The job of first responder is complex, challenging, stressful, fluid, rapidly evolving, and demanding. To add to this, we have very demanding bosses, supervisors, governing bodies, and constituents. Our clients are our own neighbors, colleagues,

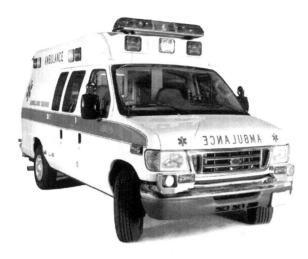

Figure 1.2 This is an ambulance. Obviously, this is what a responder would call for to transport an injured person for definitive medical care. First responders use slang terms to request certain resources on a daily basis. Generally, the dispatcher will understand exactly what resource is needed, but there are terms that do not hold the same meaning across the country. (Image copyright Leonid Smirnov 2009, under license from Shutterstock.com.)

and friends. No policemen ever want to hear the phrase *"officer down."* However, if they do hear such a transmission over their radio, you can be certain they will move heaven and earth to respond to assist their brother or sister officer in whatever manner they can. Similarly, at a structure fire, if a firefighter transmits a *"Mayday,"* all firefighters on that scene will also do whatever is necessary to get to their colleague to mitigate the crisis. There is indeed a thin blue line and a thin red line as well. Emergency service responders can relate to each other and will sacrifice all they can to ensure each other's safety. Those not involved in these fields will never understand the feeling in the pit of one's stomach when a fellow responder is in trouble. Responders need to be certain that they provide our "clients" and each other with the skills, education, training, tools, and the ability to mitigate any incident and capably handle any sudden emergencies. If responders can protect each other, they can most assuredly protect their "clients."

Does the Public Have Unrealistic Expectations of First Responders?

Another issue facing our response community is the sometimes-unrealistic expectations from those we serve. Television has truly raised the bar on what the public expects from its responders. On certain television series, forensic teams are able to solve crimes in laboratories in one hour. We see firefighters walking upright in structure fires with completely unobscured visibility and sometimes even without self contained breathing apparatus (SCBA). These programs, while certainly entertaining, provide responders with unreasonable expectations from the citizens they serve. Jurors in trials expect that crack detectives can get the murderer to confess, and if they cannot, then the defendant must be not guilty. There is an unrealistic expectation that crime scene evidence collection units can find evidence at all crime scenes, which is also not the case. The real world incidents that we respond to are not resolved in a single "episode." They are complex and often difficult or dangerous for the responders. Public expectations that are unreasonable only serve to make responders' jobs harder. Police often hear citizens mumble under their breath that "I pay your salary" and other similar comments at traffic stops. The police cannot be burdened with trying to accomplish the impossible just to impress or satisfy the needs of our citizens.

Then what is the answer? How do we serve all of our "masters"? It is not easy. Whoever is the first responder to arrive at any incident is placed under immediate scrutiny. The public not only expects, but demands that the first arriving responder begin to alleviate the problem to which they have been summoned. But what if that scenario is not plausible? Expectations sometimes lead to terrible tactical decisions. Police officers in uniform are no match for the heat, toxins, and poisons present in house fires. Why do they rush in? The noble answer is that they do so to try to save lives. The harsh and sometimes sad reality is that sometimes they react because they feel they must, or that this is what people expect them to do. Whether they are trained is of little consequence once the decision to act is made. But is this fair to these brave, if perhaps misguided souls? Do we not owe them the best opportunities to not only respond to the crisis, but also to go home to their families after their shifts? We certainly do. The question now is how to best accomplish this objective.

Figure 1.3 Most police cruisers located anywhere in America are basically the same. Some may have additional equipment such as radar units, video cameras, shotguns or rifles, but the basic components are the same. Some police units, such as SWAT vehicles, bomb response units, and K-9 units are specially equipped for their particular missions. Anyone requesting a police cruiser knows what resource is coming. (Photo courtesy of Middlesex County New Jersey Sheriff Joseph C. Spicuzzo.)

Training Is the Key

Simply put, the best way to ensure responders have the requisite abilities to begin to mitigate any incident is to provide them with all the necessary training. This training will require that the responders branch out from their traditional, day-to-day responsibilities. For example, most police officer recruits get minimal exposure to firefighting during their police academy training. But is this enough? Sadly, this will not be the case if a police officer responds to a fire and is injured due to that lack of training. Often, fire crews arrive to a working structure fire only to find a police cruiser parked in front of the burning building. That specific location of the cruiser can be bad for a tactical fire attack, but even worse is when the cruiser is parked in front of the nearest fire hydrant. Are police intentionally trying to hinder the operations of firefighters? The obvious answer is that they are not. The police are trying to get to the scene as quickly and safely as they can, but are *"situationally unaware"* of the location of the fire hydrant, or that they are blocking a critical parking place for fire apparatus. Further, police often make valiant attempts to enter burning structures to attempt life-saving activities. These efforts, while sometimes bearing fruit, are often unsuccessful, causing needless injury and illness to the law enforcement officers. Also, most police officers are unaware of the potential detrimental effects of opening a door during a working fire. This simple act, done at the wrong time or place can actually increase the intensity of the fire, or draw an existing fire to the open door, thus hindering the fire crew's entrance to the building upon their arrival.

Figure 1.4 Tanker (Fixed Wing, West Coast). These fixed wing aircraft are used to collect water or other fire suppressing agents, which they deliver directly from the aircraft to the fire. This piece of equipment on the west coast is referred to as a "tanker." On the east coast, it is called a fixed wing aircraft. (Photo courtesy of John Reith.)

Figure 1.5 This is a relatively modern, east coast fire tanker. This particular tanker carries 2,500 gallons of water and has hoses and a fire pump. Tankers will carry more water than engines and are used for suppression and support in areas where there are no fire hydrants. The equipment mounted above the rear tires is a foldable /portable tank. Water from the tanker is emptied into the tank, and through hard suction hoses, is pulled into the pump to be sent to the fire engine to be used as needed to suppress fires.

These examples are only a small sampling of how emergency responders often cross their traditional professional boundaries. In the body of this text, we will explore training deficiencies across the bulk of the emergency response community. These deficiencies will cover "routine" calls for service as well as responding to terrorist incidents. Generally, most of the efforts of responders acting outside their response discipline are not only warranted, but necessary, and the results are not always as one would hope. Given the current political and economic climate where responders are often asked to do more with less, cross-training has become a way of life. Each responder discipline has its own specified skill sets in which the responders are masters of their domain. More and more often they are required to venture into the expertise of another response discipline. Even though they have good intentions, the sad fact is that they are "flying blind" too often. Employers need to recognize that asking for responders to do more with less, in terms of compensation, training, equipment, and experience is unfair, to say the least.

Crossing into New Response Areas

In the years since the tragic events in New York, Virginia, and Pennsylvania, we have seen the phenomena of emergency responders crossing traditional boundaries on an almost daily basis. These events happen across the country, from the smallest hamlets to the largest cities in America. These responses, while always critical to the victims, have been generally small scale. But what can we expect should we experience another large-scale terrorist attack on our soil? Will our first responders be adequately trained and equipped? What if we suffer a simultaneous attack in more than one area, but in proximity to each other? Will the local incident commanders be able to implement all of their training? Will they be able to get the needed resources to all attack sites in a timely fashion to save as many lives as possible? Are all responders familiar with the principles of unified command, incident complexes, and area command? Sadly, a great many of them are not. Since we cannot be certain where the next attack will occur, we must all be ready to employ the strategies, tactics, and values associated with these incident command principles.

Many other examples can be cited to support my position that responders need better and more comprehensive training. Some of these will be explored in later chapters. A case will be made to expand training of police, fire, medical service responders, members of emergency management, National Guard, and even our political leaders. We will address issues such as the inability of responders to get along with each other outside of an emergency response area. We will try to shed some light on why this is happening. Once the problem is identified, a strategy to correct it must be created. This text will clearly identify the problems, and hopefully, several options for corrective action will be presented.

This text will delve into the recent trends of distance learning for our responders, as well as online training options. These training and educational options are some of the current hot topics in the response community. Do these programs work, or are they inefficient and not properly regulated? Each specific training session and opportunity for continuing education needs to be evaluated on its own merits.

Testing Our New Training

Once we are able to get all first responders on the same training footing, how do we test their abilities? Will the drills and costs of training render the concept of crossing traditional responder values obsolete, even before the concept is realized in the various responder communities? Are there cheaper, or even free techniques that can be employed to test and reinforce newly acquired training? There are training and reinforcement tools available to us in the response communities. We can conduct tabletop drills and follow them up with hands-on drills. The hands-on drills will certainly provide a forum to observe and test the new training areas for our responders.

Administrators versus Responder Needs

Many administrators will decry the attempts to expand training curricula for their responders for several reasons. The lack of time, competent trainers, and financing will certainly top the list. In recent years, the federal government has recognized the financial burdens of response organizations. To assist them in maintaining and expanding their training, grant programs have been developed and properly funded. Tips and suggestions for applying for and receiving grants for training will be reviewed as well. Often, agencies are tasked with beginning to enter into a new operational area of responsibility. This is a result of tighter budgets coupled with demands that responders do more. There are equipment and training grants available for agencies who take on a new operational mission. If administrators are familiar with these grants, they may be more supportive of their agencies' expanding their existing roles to include newer ones.

Are there people who are outside the traditional role of first responder, but who may be of great assistance to the first responders? There are many hard-working civilians in America who are put in positions where they can be of great service to us. But, we need to recognize who these people are, what their skill sets are, how trainable they are, and how receptive they will be to the concept of additional training options. These potential responders are located all around us each and every day. They are perfect examples of resources hidden in plain sight. We can utilize these workers to our advantage; all we need do is provide them with a little training and some encouragement.

The expression "think outside the box" has been trivialized and overused since 9/11. However, in this area of using all available resources to make us better prepared for unexpected, large-scale incidents, we truly need to explore every option. Failure to do so on our part will not be acceptable. Great numbers of hidden assets may be able to provide us with significant information if only we take the time to cultivate a relationship with them. In times of crisis or emergency, victims will take assistance from any source willing to render aid. It would be foolish of us to not seek out non-traditional sources of aid for our citizens. Americans are a very resilient and tough group of people. We have stared down adversity in the past and will be able to do it again in the future. It would be best if we were able to stare it down *before* tragedy strikes, rather than to admit defeat, and demonstrate our continued resolve yet again. We must remain vigilant and press forward to be proactive in our response capabilities and training.

These are but a few areas to be explored in the body of this text. Hopefully, by the time you finish reading the materials and opinions presented, you will understand the need to cross-train our responder communities—hearty and substantial cross-training, so as to enable our responders to protect not only our citizens, but themselves as well. The position taken in this book is that America needs to expand the traditional training of responders. They will be closely scrutinized by our citizens the next time there is any significant incident on our soil. The incident does not have to be related to terrorism and may be a result of a number of scenarios. If we demand a lot from them, then we owe them proper training and access to proper equipment. We deserve it as responders, and more importantly, the citizens expect and demand it as taxpayers. Society will expect a better effort from our responders than they have seen in the past. While every responder gives 100% effort every day, sometimes that is not enough. To be truly prepared, there still is a lot of work to be done, and a host of training that needs to be delivered. We cannot be caught unprepared again. We have seen the "worst case" scenario, and we have recovered. We need to be up to the next challenge, and do it better.

Endnotes

1. Homeland Security on the Range. Available at: www.motherjones.com/politics/2006/03homeland-security-range. March/April 2006 edition.

Chapter 2

Law Enforcement

Law enforcement in the United States comes in many different shapes and sizes. There are federal agencies that employ thousands of agents and support staff. State agencies also have from hundreds to thousands of employees. In addition to state troopers or highway patrol officers, there are state rangers and other state level agencies. Next, we get to county level law enforcement, such as county sheriff's departments or county police departments. There are tribal police agencies and local or municipal authorities as well. County agencies may also have a wide-ranging number of officers. Some municipal law enforcement agencies may be as small as a handful of officers, but may also be thousands strong, as in the case of the New York City Police, or the Los Angeles Police. Whatever the size, there are certain common denominators regarding these law enforcement agencies. Each provides a core set of basic training to recruits and transferring veterans. Given the great variety of law enforcement options available to society, it would seem that we should have much lower crime rates.

Nationally Recognized Standards Are Needed

Part of the problem in the United States is that there are no national standards governing police training as a whole. Whereas some agencies require a few weeks of training, others require months of full-time instruction. Accordingly, there is no national standard for accreditation for officers, nor is there any national certification program. For example, if I were licensed in one state as a police officer and decided to relocate to another state to continue my career, I might be required to complete my basic training again. While I will concede that there are differences in the governing and controlling statutes between states, the basic police academy training should be the same. Upon my transferring to a new state, I should be required only to become acquainted with the differing state statutes and motor vehicle laws, as well as firearms qualifications. The administrative codes will vary from state to state as well, but these subtle differences should not create a need to repeat basic training. In the military, if a soldier is transferred from one base to another, there is no need to repeat basic training. There should be a system in place to allow for a similar training curriculum for all of our law enforcement responders as well. This is a case where the military concept of training should be mirrored by the civilian authorities.

The United States government has established a program designed to protect certain urban areas from the threat posed by terrorism, determining that areas of America with the largest populations and other critical infrastructures were most at risk. They have called this program the Secure the Cities Initiative. The chosen regions have been awarded millions of dollars in grant funding to "secure" or harden their areas. In addition to target hardening and physical tools being purchased, training funding was also made available. This training was extended to all types of first responders, and certain programs have been developed to assist in getting responders trained. Law enforcement resources have been made a part of this training in these areas. The areas are referred to as Urban Area Security Initiative (UASI) regions. Each region has established a law enforcement reactionary force to be trained and utilized under prescribed criteria. These officers have received training above and beyond what other responders in these UASI areas receive if they are not a part of that reactionary force. It would seem that the remaining officers in these areas should get the training as well. Imagine that the reactionary force is deployed to an area in the UASI Region. While they are away, an attack occurs in another, unprotected part of the region. This would then leave the targeted area with a number of officers who are lacking certain training. While it is understood that grant funding has its limits, there needs to be a better way to distribute the training. Perhaps train-the-trainer programs can be developed, and the training can be given to all of the members of these departments. But what if we guessed wrong and the terrorist attack is not in a UASI Region? Should we not make an effort to train all of our law enforcement officers for such a scenario? This training would certainly be costly, but would it not be worth it? Training should not be restricted to certain areas of the country to the detriment of others.

It would make sense to me that responding to a residence for a theft report would be the same in New Hampshire as Montana, but this may not be the case. I would imagine that the tactics involved in a motor vehicle stop would be universal. After all, we have seen all too many law enforcement officers killed during such "routine" activities. However, this is simply not the case. Some states allow officers to shoot at tires of vehicles, while others do not. Another difference is that domestic violence laws vary from state to state. Certain states share a fundamental basic training philosophy with each other. These states will recognize the training provided by each other in cases of transfer. However, this "reciprocal training agreement" is actually quite rare. Most states will, as already indicated, require completion of a full police recruit academy program prior to licensing its new officers. Prior training and experience are of little matter to the hiring state and agency. While there is a tenuous and fragile consensus on certain strategies and tactics among all law enforcement agencies, a full training program is required.

NIMS Attempts To Address Certain Issues

How can we justify a reciprocal training agreement between some states but not others? This very question is one that the National Incident Management System (NIMS) is struggling with. Those who crafted the NIMS program realize that this very question is in need of serious examination. For all states, there should be a national basic training standard. Whether local, county, or state agencies, a national standard needs to be established to create one set of training to which all new recruits are exposed. Regardless of the size or jurisdiction of your agency, recruits in all states should enjoy the same core training. Federal agencies as well should adopt a basic training program for all federal agencies. After basic training, all states would then have the ability to add additional training that would be specific to the laws and requirements of their state. In this manner, all law enforcement at the local, county, and state level would share a core foundation for future training.

It is unacceptable in this age of technology that we cannot agree upon a standard number of training hours for all of our law enforcement recruits. Such a program may be difficult to initiate, but in the end, it will pay huge dividends. Generally, there are those desirous of a law enforcement career that will get an entry-level position to obtain basic training. Once trained, and after gaining some practical experience, these officers look to move on to "greener pastures" in other agencies. These lateral moves are sometimes motivated by financial reward, hours and types of shifts worked, promotional opportunities, and specialty opportunities in agencies with specialized units such as K-9, bomb disposal, etc. When one of these officers seeks to transfer across state lines, the training process starts all over again. Having a standard core police academy can allow for a much quicker turnaround for new transferring officers. This will get the "new" officer on the beat much faster.

Basic or entry level law enforcement training is just that—basic. New recruits are put through their paces in a wide variety of topics. Included in them are state law, firearms, traffic laws, report writing, domestic violence, building sweeps, certain tactics, self-defense, physical training ethics, first aid and CPR, and chain of command. Many police academies have a small block on fire response and basic hazardous materials response. In my case, my recruit class had a fire engine come to the police academy, and a firefighter showed us some of their equipment. While it was over twenty years ago, my recollection is that we spent less than one hour with this firefighter. Later in the academy, we went to a local fire training facility. A small fire was lit in the training area, and we were made to crawl through a maze under mild smoke conditions. We did so without the benefit of any respiratory protection, and were supervised by what we assumed were fire service instructors. The purpose of the exercise was to allow us minor exposure to smoke-filled environments to give us some perspective on what a real fire might be like. While many of my recruit classmates were nervous prior to this drill, my prior fire training assisted in having a calming effect on them. In fact, most of my classmates wanted to be my partner for the drill since I was very confident that I could manage the exercise. That was the sum total of my "cross-training" relating to the fire service while in the police academy.

Law Enforcement Terminology

One area in which we can mostly agree in law enforcement is in the area of language or terminology. While there are some minor differences, we should not have the same issues that the fire community faces. The names police cars, squad cars, and cruisers are all used, but they do all refer to the same vehicle. Aside from some specialty weapons systems carried by some law enforcement, police cars are all basically the same. They are generally sedans or four-wheel-drive vehicles; most have protective cages, radios, emergency warning equipment, possibly computers or shotguns, a first aid kit, and maybe a semi automatic external defibrillator (SAED). Police cars are not *typed,* which we will address later in this text. Basically, if law enforcement were using plain language, there would be little confusion as to which resource they were referring.

The law enforcement community still uses ten codes on a daily basis. There is some debate about this. The debate stems from the mandate from NIMS that all responders use plain language on the radio and during communications. In 2006, the Chiefs of Police Association asked then-Homeland Security Secretary Michael Chertoff to reconsider this provision of NIMS. The position of the police chiefs was that they needed to use these codes in-house to keep airwaves free for emergency transmissions and to provide for some security among police. Secretary Chertoff agreed with the chiefs and allowed police and other agencies to continue use of the ten codes on a daily basis. He did, however, mandate that if more than one agency was involved in any radio communication, plain language must be used. There was uproar from some in the response community, alluding to the fact that Secretary Chertoff backed off from the NIMS guideline under pressure from law enforcement. In fact, Secretary Chertoff did not, and he listened to the arguments put forth by the chiefs. We do train law enforcement personnel to understand that once they are working with anyone else, plain language must be used. With the regular availability of police and fire scanners to the public, it would seem that plain language would be better used. However, unlike most other response disciplines, there are ten codes that are still used that do save time and free up *"talk groups"* for other responders.

Still, there are some issues with terminology among responders in various response communities. Firefighters and hazardous materials technicians use different terms for exposed areas at an incident. For example, when fire or hazmat arrives at an incident, we will establish *"control zones."* These are broken into three zones.

First is the hot zone, where product may be released, or where the tactical operations are being conducted. The next zone, working away from the incident toward a safe area, is called the *warm zone*. This is also referred to as the "contamination reduction corridor." This is where we conduct decontamination and undress responders who were exposed or contaminated in the hot zone. The last, farthest area from the operations, is the *cold zone*. This is a completely safe area, where no personal protective equipment (PPE) is required.

Law enforcement has a similar system, but they use different terminology. They use the term *perimeters*. Law enforcement personnel will establish inner and outer perimeters, which are similar to the fire and hazmat control zones. This is a subtle difference, but if police and fire or hazmat interact, we must all have a similar understanding of what we are talking about.

There is one last note in this area that may be of some significance. When firefighters arrive at a multifloor structure, we number the floors from the bottom up. I was speaking with the commander of our local Special Operations Response Team (SORT, our name for what is commonly called SWAT), who advised me that their personnel number the floors from the top down, which is totally opposite from the way fire and hazmat are trained. We discussed the matter, and he indicated he understood why fire did things opposite of SORT and agreed to change their system to match fire's. This may not seem important, but with highrise buildings, if fire is working in one direction and law enforcement SORT in the opposite, we can miss our intended event by many floors. Further, the fire service refers to the floors in a building as *divisions*, while law enforcement calls them floors. Fire service personnel will designate sides of the building by letters, such as A side, B side, and police by use of direction, such as east or west, another area where we can be looking at the same event and speaking two completely different languages. We really need to consider these deficiencies in our communications and make some decisions before we pay a tragic price.

In the late 1980s, when I was a police recruit, we received no significant hazardous materials training in the police academy. We were, however, required to complete a thorough course of instruction in the area of first aid and CPR. One can reasonably deduce that the reason we were given a thorough first aid background is that we can expect to be put in a position where we may be required to administer first aid or CPR. Back then, there were no SAEDs, though now this training is given at the recruit level. It seems that it was anticipated that law enforcement would be first to arrive at medical emergencies, and we trained our recruits accordingly. However, those in charge of the various law enforcement academies were not so insightful regarding hazardous materials response or fire emergencies. Common sense would seem to dictate that police already on the road, always on duty, with their knowledge of their geographic area of responsibility and without the need to don additional protective garments, would be the first to arrive at any emergency.

Law enforcement in the twenty-first century is more dangerous and stressful than at any other time in our history. Law enforcement professionals are now

Table 2.1 Partial Listing of Chemical Agents

Chemical	Type	Odor
Sarin	Nerve agent	None if pure
Soman	Nerve agent	None/camphor
Tabun	Nerve agent	Faint fruity
VX	Nerve agent	None/sulfur
Mustard gas	Blister agent	Garlic
Lewisite	Blister agent	Geraniums
Chlorine	Pulmonary	Ammonia type
Pepper spray	Irritant	None distinct
Cyanide	Pulmonary	Almonds

required to be bilingual in many cities and towns, are exposed to violent street gangs even in small-town America, respond to fire emergencies and serious and sometimes violent motor vehicle accidents, and, in addition, now must factor in the possibility of terrorist attacks. Years ago, there were many more two-man patrol units on the road. Financial issues and constraints have removed the second officer from many cruisers. This makes backup farther away and exposes us to dangers for a longer period of time on even the most routine of civilian contacts. Regarding the fire aspect, the fuels that support combustion are filled with poisons and toxins, which can lead to cancer or other respiratory conditions. With the medical response issues, current officers are faced with a variety of bloodborne and airborne pathogens. Additionally, existing Health Insurance Protection and Portability Act (HIPPA) laws make it difficult to determine whether responders have been exposed to any of these deadly viruses or diseases (Table 2.1).

Given all of these factors, one would think there would be a dearth of qualified police applicants. In fact, in the economic climate of twenty-first century America, a career offering a pension and health benefits with little fear of loss of job is a very appealing prospect. Many soldiers returning from service abroad see law enforcement and firefighting as viable careers. Most police officers at any jurisdictional level are Type A personalities. They are self-motivated, aggressive people who desire to help those in need.

Helping Our Law Enforcement Responders

While the pool of applicants is very qualified and deep, are we truly putting these officers in the best possible position to succeed? Are we doing all we can to ensure their safety? We as citizens call upon police for a number of issues that we ourselves cannot control. Once they arrive on location, it is expected that the officer or officers will be able to put our fears to rest, provide a sense of security, or simply refer us to the correct agency to deal with our problem. They are our problem solvers. Law enforcement officers are the first responders that most Americans have interaction or experience with, whether it be their coming to our schools as resource officers or seeing them at career fairs. We also see their patrol cars in our communities. Some civilians have contact after having been stopped for traffic infractions. The lucky are given warnings, while others receive a written memento of the contact with the officer. Children are taught to seek out uniformed police if they are in trouble. We arm these officers and give them authority to take away our freedom, and in certain situations, we authorize them to use deadly force to preserve human life threatened by criminals. We expect and demand a lot from these dedicated professionals. If we expect a lot from them, they should expect a lot from us. We all know that in our society today respect for police authority is no longer a given. In fact, the total disregard for human life is all too evident in our culture. Every year, close to 200 law enforcement officers will lose their lives in the line of duty. This is a totally unacceptable number and ongoing trend, especially as most line-of-duty deaths are senseless. Some officers will be killed in their own cruisers responding to calls for assistance, some by drunk drivers, some while sitting on the side of the road. Some will die as a result of being struck by cars, some stabbed, some shot. In an unusually cruel scenario, some law enforcement officers will be ambushed, and, quite simply, murdered in cold blood. Whatever the manner of death, we can expect far too many officers to die each year.

Unrealistic Expectations Can Be Deadly

All recruit level officers know that there may come a time when they may be called upon to use deadly force. They also know that this same deadly force may be used against them. Sadly, too many officers perish each year. While there are generally no *acceptable* line-of-duty deaths, what is totally unacceptable is when officers perish doing duties for which they have not been properly trained. These officers are doing what they think is the "right" thing to do. Often, this false sense of duty turns deadly. As "problem solvers," they feel an obligation to do something, even when they are not properly trained to act. For example, if officers respond to a motor vehicle accident where a power pole has been compromised, and there is a live wire lying across the car, there is very little they can do. Even my recruit police class was trained in this specific scenario. The live wire has energized the vehicle, and to attempt to enter the vehicle, or even to touch the exterior of the car could lead to death by electrocution. With no apparent or claimed injuries by the driver or occupants of the vehicle, law enforcement officers will be satisfied to await power crews to cut the power to the energized lines.

If the scenario changes, however, the actions, which should remain the same, will also be altered. If we add the fact that the driver has suffered a stroke, heart attack, diabetic emergency, asthma attack, or any other such medical emergency, the officer knows that the driver may perish without medical intervention. The officer has called for an expedited response by the local power company. Does the medical crisis on the part of the driver change the facts regarding the vehicle's being energized? Obviously, it does not. While there may be a desire to act, it will probably be resisted by the officer. Now let's change our accident scenario again to include the fact that one of the victims is a young child. The mother is begging the officer to do something. A crowd has gathered at the accident scene, and they are imploring the officer to intercede. Will the officer do anything? He or she may very well have a strong desire to act, but doing so will probably mean death. There have been cases where bravery replaced common sense, and officers have been killed trying to render aid in situations where immediate intervention was not possible. We have trained our police (and firefighters) that energized power lines and human beings are incompatible, yet even with protocols in place, some responders are determined to tempt fate. Sadly, many of those who choose to roll the dice in these situations suffer serious or mortal wounds or injuries.

Why Police and Fires Sometimes Do Not Mix

A second real life example to consider is one that also happens all too frequently. A structure fire is reported, and it is unknown if all residents have escaped the blaze. The police department/sheriff's department is the first to arrive. Their focus is on the fire and the possibility of a trapped victim. Many times, officers are not "situationally aware," which, in this case, means they either block a fire hydrant or stop in front of the burning building, possibly being ignorant of the tactical needs of the fire crews. Fire officers know that the ladder or truck company needs to park in front of the building. The fire engine needs to access the fire hydrant. The well-meaning police officers have severely affected the fire response, and the resulting delays may make rescue and suppression efforts significantly more difficult.

While awaiting the fire responders' arrival, the neighbors indicate there is an elderly man who resides in the residence, and they have been unable to locate him outside his burning home. Seeing the smoke from the fire begin to intensify, police officers make a decision that to wait any longer may result in the death of the man. So the officers enter the building in search of the man. The law enforcement responders are afforded no protection from the heat in the form of firefighting clothing, nor are their lungs protected from the heat, toxins, and poisons present in the fire. Sometimes the officers will retreat before it is too late, and will suffer only from smoke inhalation. Sadly, the by-products of the fire too may overcome some officers, and they become victims for the fire crews to locate and retrieve. Given all of these facts, why do police professionals enter buildings that are on fire, knowing full well that they have no protective gear? Because the public expects them to, and in many cases, they feel provoked into doing so. The officers are trained to know better, but sometimes confuse bravery with common sense.

Partially Trained Responders Can Be Easily Hurt Or Killed

There is a common theme in both of the aforementioned scenarios. The officers act, imprudently and against their training, to try to effect a better outcome for the victim. The other similarity is that they have been trained to know better, but acted anyway. Common sense alone should have prevented actions in both scenarios. However, they ignore their training and life experiences in an ill-fated attempt to assist a total stranger. I do not believe, as some do, that officers disregard their safety and training for glory, accolades, or medals. They do so out of a genuine desire to help the victims. The common factors are that they have had *some* training, but acted in accordance with their belief that they would be in a position to bring about a positive resolution to the incident or emergency.

The NIMS program has included in its training the concept of categorizing and quantifying resources. This system has two basic components, known as "kind" and "type." What the system refers to specifically is which specific resource is needed (kind) and what is the capability, or performance, of that resource (type). For example, in law enforcement, there are a group of resources that may be called to specific incidents, but that need to be capable of certain tasks. If police respond to a missing child report, they may decide to call for a K-9 team. This would be the *kind*. However, a more specific request would be needed, such as a tracking dog *(type)*. Simply asking for a K-9 team without specifying which type may result in getting the wrong resource. Certainly, an explosive material-detecting team, or a drug-detection team would not assist in tracking a missing child. This concept of kind and type is already used in a practical sense every day, but under NIMS, many resources will be typed or classified for use in the field. In this manner, all responders in every state would be referring to the same resource with the same terminology. Currently, as previously noted, this is not the case.

Responses Vary between Response Organizations

Police officers spend a great deal of their workday patrolling their response areas in their cruisers. In today's society, we see hazardous materials tank trucks on our surface roadways everywhere. These trucks drive down Main Street USA, on interstate highways and intrastate roadways. Often they are carrying very toxic substances and materials. Should one of these trucks be involved in an accident, who will be the first arriving responders? The police are generally the first to arrive. But specifically, what are they trained to do in response to an incident involving these hazardous materials carriers? Their training is to use the shoulder of the road and get as close to the accident as possible to determine the nature of the problem, then request appropriate backup and response from other agencies. They then exit their cruiser, and begin a tactical assessment of their scene. They request fire, ambulance(s), hazmat teams, tow trucks, etc. But what if the truck is leaking highly toxic or poisonous materials? What would happen if a breath or two of a chemical or material released is enough to be fatal? Then all they get to be is *dead*. This is why firefighters jokingly refer to police as *"blue canaries"* or *"copological indicators"* when responding to hazardous materials calls. The joke among the firefighters and hazardous materials technicians is that they can safely advance toward the scene at least until the last standing police officer.

Firefighters are trained to always approach from an upwind direction if at all possible. This provides the safest path and the best chance of survival and safe operations on-scene. Police do not have the same training and will approach the scene as quickly as possible without due deference to many other factors that may adversely affect their health and safety. Firefighters are equipped with binoculars and are taught to stop a great distance away. We are then taught to use the binoculars to gather information about the load being carried by the truck by reading the Department of Transportation (DOT) placard. Fire officers then use the North American Emergency Response Guidebook (NAERG) to identify the product involved. This book, provided to emergency response organizations at no cost, is a small reference tool used by first responding personnel.

NAERG consists of four main components. The chemicals that are mentioned are represented in alphabetical order in the first section. In the second section, they are listed by the number shown on their placard. The placard is a diamond-shaped labeling system placed on a vehicle so that it is visible on all four sides of vehicle. All chemicals represented in the guidebook are broken into classes, such as explosives, flammable liquids, corrosives, etc. Some of these classes are broken down further and are so identified in the front section of the book, which lists the classes as well as a photo of the placard for each class. The third section of the book contains the "guides" for each chemical listed in the body of the book. A guide may be used for a group of chemicals, and accordingly, there are far fewer guides than chemicals detailed in the book. The guides are orange-colored pages that provide the responder with pertinent information about the chemical, such as fire suppression tips, first aid instructions, etc. The last main section of the guidebook provides evacuation distances for day and night releases that are either large or small. Not all chemicals will be represented in this section. Users of NAERG can then refer to a specific guide in the book that will advise how to best handle the scene. Later, if prudent, crews will move closer, don PPE and self-contained breathing apparatus (SCBA) and begin *defensive* operations. Many of those steps are excluded from the police officers' protocol. Do police even carry binoculars in their cruisers? Many agencies do not place them in their vehicles, and this may have serious consequences when that tool is needed by the officers.

Obvious Warnings Are Not So Obvious to All Responders

The National Fire Protection Association (NFPA) has a four-diamond system that it uses to alert responders to certain aspects of hazardous materials. This system is known as the NFPA 704. This is a large color-coded diamond that is broken down into four smaller diamonds colored blue, red, yellow, and white. This system is not a placard, but rather a label system. Labels are generally placed on boxes, as opposed to placards, which are placed on vehicles. Labels are also used on fixed storage containers such as may be found in chemical storage facilities. In this system, each color corresponds to a different aspect of each substance. The range of numbers for blue, red, and yellow is from zero to four. Blue refers to the health effects of the substance. A material with a zero would be relatively harmless, while a blue number four would be very dangerous. The red diamond is in place for flammability. Therefore, a red diamond with a zero means the substance will not burn, and a red four means that the material has a low flashpoint and will ignite much more readily. The yellow diamond is set for reactivity. A yellow diamond with a zero means that the material is very stable, while a yellow four means that it is very unstable and volatile. The white diamond is reserved for special features of the material, and symbols and words are used, rather than numbers. A letter "W" with a line through it warns the responder not to use water. The other primary word is "oxidizer," reflected as OX. Occasionally, other words or symbols such as the radiation trefoil, COR (corrosive) or BIO (biohazard) may be used as well. This system is in place to provide responders with some information about the particular material. Most law enforcement personnel are unaware of this system, and if they saw the label, they would probably not understand the meaning. Again, this is very important information that is available, but most law enforcement responders do not have the training to understand exactly what the system is telling them.

Would officers' responses and actions be different if we gave them more training regarding this particular materials characteristics system? It would certainly be of great benefit to law enforcement responders to have an understanding of this labeling system at an actual incident. The more training responders have, the safer they will be. We need to be sure that all responders see and interpret any warning labels in the same manner.

Sometimes Doing Nothing Is Better Than Doing Something Wrong

There are other examples where lack of training can prove deadly to emergency responders. Police are dispatched to a report of a possible fire in a three-story row home. They are first to arrive on the scene and observe a moderately smoky environment. They kick in the front door and begin knocking on the upper-floor doors to alert residents to the fire. Unknown to the officers, there was a smoldering, oxygen-starved fire in one of the apartments on the second floor. The officers have no real fire training. While this lack of training may not be their fault, in this case, it will have deadly consequences. While they are stirring residents on the third floor, the fresh air, including oxygen, was introduced into the environment, and the smoldering fire has flared up. Before they can get down the stairs from the third floor, the fire has extended to the stairway on the second floor. Now the officers and the residents from the third floor are trapped, and the stairs are ablaze. Had the officers closed the door after entry, the fire would not have been refreshed by the introduction of fresh air, and it would not have been as serious. In this case, the officers are not as wrong as in the previous two scenarios. Why are they not wrong? They are not fully trained about the effects of improper ventilation or the effects of air on oxygen-starved fires. This is a specialized aspect of firefighting that is not commonly understood outside of the firefighting community. All three incidents described are serious, but in the last one, the officers had no training to rely on to ensure their safety.

Do All Responders Understand the World We Live In?

Continuing in the fire or hazardous materials training, how often are police dispatched in advance of fire service personnel for carbon monoxide alarms, or for persons taken ill, or for fire alarm activations? Carbon monoxide is a gas that has no odor, color, or taste. It is invisible and not detectable by humans without tools specifically designed to detect the presence of certain gases. Carbon monoxide kills hundreds of Americans each year, mostly while they sleep. However, if there are great concentrations of carbon monoxide present in the air, such as 400 parts per million (ppm), an officer can quickly become overwhelmed, and could become disoriented, lose certain motor skills and eventually pass out. Carbon monoxide in an enclosed area has the effect of displacing oxygen from the air. The normal concentration of oxygen in the air we breathe is only 20.9%. A drop to 19.5% is indicative of a low oxygen concentration. Officers exposed to an oxygen-deficient environment without respiratory protection can become disoriented and eventually lose consciousness. If they are not removed and treated in a timely manner, death can occur. Even though this issue is very serious and prevalent in many areas, most police officers receive no training in responding to carbon monoxide alarms and do not even know at what levels we must evacuate residences. In many fire departments, we must evacuate the residence at 35 parts per million. A reading with as little as 8 ppm of carbon monoxide will activate a request from the gas company. Police need better (or at least some) training regarding this potentially fatal call for service.

I have already shown where lack of training can be dangerous for the law enforcement community, as well as for the public they serve. There is yet another area where training is missing, and a lack of understanding of consequences by law enforcement can have dire results. The concept of control zones, which police often refer to as perimeters, is another such area. If there is a hazardous materials incident or a terrorist event, and victims have been exposed to either hazardous materials or substances or chemical warfare agents (CWA), they will need to be decontaminated prior to receiving medical treatment or transport to a medical facility. Our emergency medical responders are very well aware of the need to decontaminate all victims prior to treatment and eventual transport. When I conduct training of police and other law enforcement personnel, I am very surprised to learn that they are unaware of this process. Even if the injuries are life threatening, if the victim has been exposed to certain agents or chemicals, they *must* be decontaminated to avoid contaminating the responders or the transport vehicle. Untrained law enforcement responders directing treatment of contaminated victims may have the undesired effect of contaminating other responders and resources. This type of training is generally overlooked and would not require great amounts of time and effort to be presented to the officers.

We have identified areas where officers are trained, yet disregard their training and are injured or killed. We have also seen incidents where officers proceed with actions where they are untrained and are subsequently injured or killed. So how do we address these issues? We need to start providing training across traditional responder boundaries. We need to think outside the traditional box and prepare our responders for what they may have to deal with. What training do we give police for responding to acts of terrorism? While the City of New York Police Department has many specialty units, what additional training is provided to the beat cop? Were they ready or even capable of any meaningful response on 9/11? What about the police near the Pentagon? Were they similarly caught off guard? We expected those officers to provide assistance to the injured, dying, and trapped, but what did we do to prepare them for this? Sadly, we probably provided no or little training. Since then, we have pumped billions of dollars into training and equipment to deter terrorism, and to respond to attacks. Some cities and areas are prepared, but most of America's front line of defense is untrained, and not properly equipped. We need to do better in these areas, and in fact, we *must* do a better job.

Suggested Solutions

In reviewing this problem, one can conclude that there is no fast and easy solution. There are a great many solutions, but none are easily implemented or achieved. In fact, the obvious first question is where do we begin the process of implementing additional training and obtaining the equipment and tools needed to achieve this goal? There is a need to outline a training curriculum, one that satisfies the broad cross-training needs of our law enforcement community. Once this curriculum is determined, a decision on presenting the material needs to be developed or chosen. The training language and material need to be comprehensive, yet understandable. There is also a need for the materials presented to have a high degree of currency, so as to best prepare officers for what they may encounter. There must be a way to get the rank and file to buy into the new area and training topics, especially the veteran personnel, who may see no value in new areas of training. As the saying goes, you can't teach old dog new tricks. We need to convince the grizzled veterans of the force that they need this training, and they have to embrace it. We need a realistic timetable for implementation. This new training needs to become mandated and added to the curriculum at police training academies throughout America.

As we graduate the next generation of law enforcement officers, a concerted effort must be undertaken to ensure that they have access to all of the available training classes. The current trend where police personnel are being asked to perform more and varied duties can be expected to continue into the foreseeable future. Training needs to be provided if this demand for delving into new response areas is ongoing. Handling of calls that were generally fielded by other first responders is here to stay, and our law enforcement professionals need to be trained and mentally ready to answer these service calls.

Additionally, we need to convince our law enforcement leaders and the politicians in control of our budgets that we need this training. Chiefs, sheriffs and public safety directors will be reluctant to outlay the financial and manpower resources for training that is not mandated. After all, they have other more pressing issues with which they are involved. There are gangs, murders, drug issues, burglaries, white-collar crimes, and other issues that are prevalent and in need of attention as well. As a society, we have become complacent since 9/11. There have been no other terrorist attacks on American soil from international groups such as Al Qaeda. Accordingly, the priority that was present in 2001 is now a distant memory. We need to be certain about one thing: Our foreign extremist enemies have not abandoned their plans to attack us. History has shown that they are, if nothing else, patient. We must not turn our backs and a blind eye to these fanatics and their agenda to kill our citizens and first responders. We need to remind and convince our leaders that the threat of a terrorist attack still remains, even with the passage of time. We need to act on that notion, and begin preparing our police officers and other law enforcement professionals for an act that everyone hopes will never happen, but may very well occur. Emergency service organizations are like a public safety insurance policy. We pay taxes for the service, but we all hope to never need a fire department, ambulance, or police response to our homes. In our personal lives, we carry insurance for our cars and homes, and yet we hope to never need to file a claim. Then why do we carry the policy? We carry it "just in case" something happens. The public "carries" us as insurance for the same reason. Just as we would expect our insurance carrier to honor a claim, the citizens have the same expectations. We need our bosses to realize this as well.

Learning from Each Other

Many different examples of training and principles are observed by one response discipline and not another. An example of this would be an accountability system to track first responders. I am not referring to a system wherein a person is tracked for an entire shift minute by minute, second by second. Rather, accountability refers to a supervisor's being made aware of responders' locations or actions in a general sense while they are working at an incident. The fire service has been using an "accountability" system for decades. The system has different variations, but the premise and goal are the same. Fire crews carry "tags" on their gear (Figures 2.1–2.5).

Figure 2.1 "Hard" accountability tag used by the author. These or similar tags are used daily by fire departments to assist in tracking firefighters at service calls. This simple system can be used by other response agencies for the same purpose. These tags are surrendered to a safety officer so the member can be tracked while inside a structure fire or similar service call.

Personnel Accountability Tag

HELMETTA

FIRE DEPARTMENT
Helmetta, New Jersey 08828
EMERGENCY MEDICAL INFORMATION INSIDE

Figure 2.2 The front outside of the soft accountability tag identifies the name of the agency of the member wearing the tag. This tag would be kept with the responder.

Different systems require anywhere from two to four tags. Once a fire call is received, one tag is placed on an accountability ring in each piece of responding apparatus. Upon arrival at the scene, the ring is given to either the safety officer, accountability officer, or incident commander, depending on either the manpower at the scene or the system in place. If the call for service is a structure fire, upon entering the

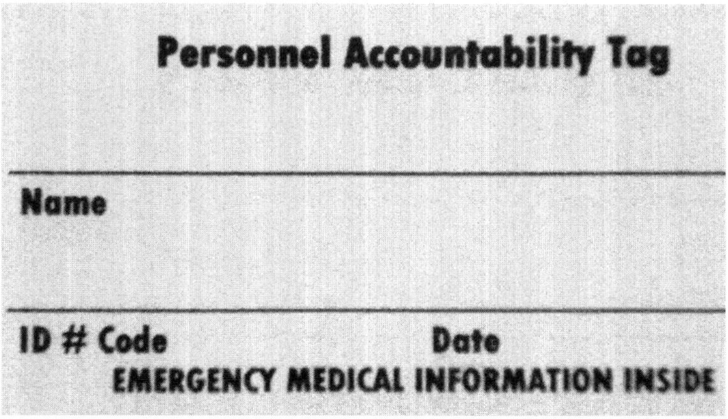

Personnel Accountability Tag

Name

ID # Code   Date

EMERGENCY MEDICAL INFORMATION INSIDE

Figure 2.3 The outside back of the soft accountability tag contains the name, identification code and date the tag was created.

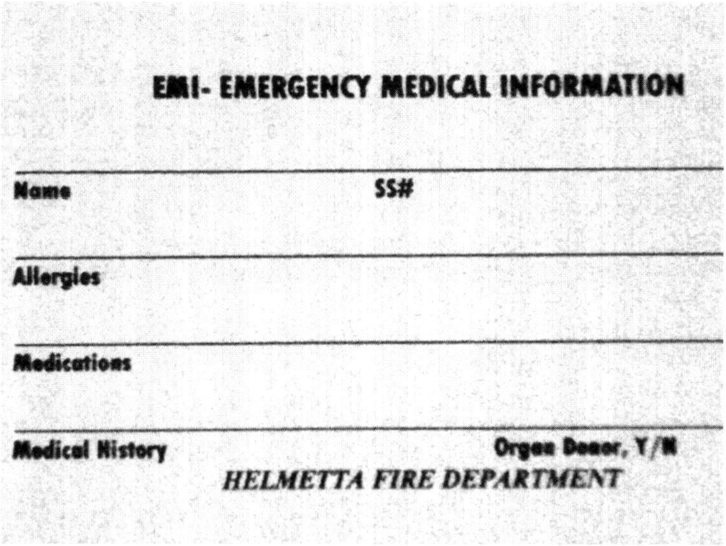

EMI- EMERGENCY MEDICAL INFORMATION

Name   SS#

Allergies

Medications

Medical History   Organ Donor, Y/N

HELMETTA FIRE DEPARTMENT

Figure 2.4 The inside left-hand side of the soft accountability tag records the responder's name, social security number, allergies, medications, medical history and whether he or she is an organ donor.

building a second tag is either handed to a safety officer or placed in a pile near the point of entry. In this manner, should there be an emergency at the scene—perhaps a collapse—that requires an evacuation, a quick personnel accountability report (PAR) can be taken. One would simply collect all of the tags on the scene and conduct a roll call until such time as all responders are accounted for. If there is a

Religion:_____

B/P (Base Line)		Pulse Rate

Date of Birth	Blood Type	Sex

Emergency Contact Person		Phone #

Physicians Name		Phone #

Figure 2.5   Inside right side of a soft accountability tag contains the religion, base line blood pressure and pulse, blood type, sex, emergency contact person and phone number, and the name and phone number of the responder's physician. If a responder is unable to answer questions, medical personnel will cut open this soft accountability tag and will have access to basic patient information. In rare cases, this information can have a great impact on the treatment and recovery of responders.

responder missing, we will quickly know who it is. Once the identity of the missing firefighter is determined, that particular responder can be tracked based upon his or her assignment and last known location. Rescue teams can be dispatched to the last known location of the missing firefighter to locate and remove him or her to safety.

In many systems, the last tag is kept on the person of the responders. Should they suffer a medical emergency and be unable to answer questions, this last tag is used. It is generally a "soft" tag, heavy-gauge paper or cardboard, which is folded in half, then laminated to protect the contents. The inside portion contains a medical history, allergies, conditions, medications, etc. An unresponsive responder will have the tag cut open, and medical treatment personnel are made aware of the responder's medical needs. Law enforcement has no such system in place.

While some agencies are able to track officers via GPS systems in the cruisers, there is no strict "accountability" system in place. Very often, police officers "self-dispatch" themselves to another officer's calls for service. There are a variety of reasons for why this happens. Boredom, curiosity, or a need to ensure a fellow officer's safety are but a few reasons one would self-dispatch. However, this practice can be a bad tactic as well. Perhaps you have seen this phenomenon playing out on our streets. You will see a traffic stop with three or more cruisers. As you go by, you see that the driver is an elderly person. Did we really need that many officers to safely handle an elderly person at a car stop? Probably not, but this activity occurs regularly. When can self-dispatching turn deadly? I will refer to New York City on 9/11. Their dispatchers did not send many of the police officers who responded to the incident location. Many officers saw and heard the events unfold and decided to respond to the scene to offer assistance. Many of these officers did so without notifying the dispatchers as to their exact location, or as to what they were doing. Once the buildings collapsed, police supervisors were unaware that many of their colleagues had gone to assist at the incident. Did these officers have a desire to respond without being "accountable?" Probably not, but you can see if we did not know who was there, it would become increasingly difficult to account for all of our responders.

Federal Assistance Used

Given all of the examples provided to date, there is an obvious need to cross-train our law enforcement community. After the attacks on 9/11, there was an effort to equip our first responding law enforcement responders with the tools that would be needed to combat terrorism and respond to any future incidents. Federal dollars were made available to purchase air purifying respirators (APR), PPE such as chemical protective ensembles, vehicles, body armor and a host of other tools and equipment. Many law enforcement agencies took advantage of the federal generosity and made purchases for their personnel. But, did these purchases include training the police on the equipment, the applications of the equipment, and the liabilities of the equipment? Was maintenance and continuing training contemplated for the officers and their new equipment? Policies and procedures need to be created or amended when new equipment is purchased. Responders need to have the training prior to being given the new equipment and placing it in service for use in the field. If officers are given equipment they had never seen, a thorough training regimen needs to be developed and implemented. Having equipment without the understanding of its limitations, uses and care serves no purpose. If responders do not feel comfortable with a new piece of equipment, they will not use it.

Simply purchasing equipment because it is available is of little value. I have seen equipment purchased for homeland security applications that were distributed without mandatory follow-up training. For example, APR are excellent tools for protection of the lungs of the user in cases of certain chemical agent releases. Prior to using the APR, personnel must be fitted to ensure the mask achieves a good "seal" around the face of the user. The user must have a machine provide a *quantitative fit test*, which will ensure that the APR properly fits the face of the user and functions within certain predetermined parameters. Additionally, the user is required to receive medical clearance prior to the training, and the training needs to be repeated annually for each user. Existing federal Occupational Safety and Health Administration (OSHA) standards also provide that the fit test must be completed for each user every year. Many agencies were not fully aware of the training and testing requirements for this equipment, and the APR has either not been issued or the officers are out of date regarding their training. Having untrained personnel or equipment that remains in storage provides no protection for the officers or the citizens.

Further, certain chemical agents can expose people to their effects if the chemical is *absorbed* into the body through the skin. Law enforcement personnel were given APR, but not all were given the chemical protective ensembles to protect their exposed skin. Many law enforcement officers may not realize that they could be exposed to effects of the agents through their skin, even if they have donned their APR. Providing equipment must also include having all of the required personnel fully trained on its use.

Police and sheriff's personnel respond to a wide assortment of calls for service every day. The "routine" calls become predictable, and, therefore, we can handle them rather easily. But what of a major terrorist attack? Are officers familiar with attack methodologies? Do they carry any PPE to ensure their safety? Do they have access to APR to maintain clean airways? Sadly, most do not. We need to train our police how to react to these types of responses. Officers need to have the ability and skill sets to protect themselves in order to be in a position to aid victims in need of assistance. Agencies provide basic and in-service training regarding many physical skills, including firearm proficiency and defensive tactics. We need to include training for these officers so that they are able to observe and react to signs and symptoms of all types of terrorist incidents. But how can they be trained? A myriad of training classes are available, and administrators at all levels need to know which courses are good, and which are not. There are certain core classes that all first responders should have, but do not. They include training in chemical, biological, radiological, nuclear and explosive (CBRNE) response, hazardous materials response, incident command, radiological and explosive specific training, first aid, counterterrorism awareness and classes in responding to terrorism.

Training Law Enforcement Personnel Early in Their Careers

Many of these classes can be provided to law enforcement personnel at the basic academy level. However, even if this is implemented, we will still have thousands of older, veteran responders who will be in need of this training. The problem with this type of scenario is that older officers may be reluctant to attend training in new areas. They may think that they have not needed it to date, and, therefore, will not need it now. We must convince all police officers, no matter their location, agency type or size, that they need this training. Once they see the need, they could possibly demand that it be provided. We will discuss some different options such as grant availability, mutual aid or memoranda of agreement as training tools later in this book. Suffice it to say that the initial concern is agreeing that there is a real identifiable need for the training. Next, a set curriculum needs to be decided upon and implemented. This will involve issues such as scheduling, training materials, compensation for attendees, arranging for instructors and payment of same.

Incident Management

Whenever there is a large-scale incident, there is an obvious need for it to be "managed." In the United States, we manage these, and in fact, every incident, using a proven management system. Emergency responders have used this system since the 1970s. At that time, those in charge of managing the wildland fires, which burn hundreds of thousands of acres, realized that they were in need of a better system to oversee all the manpower and resources operating at these fires. The system is called the incident command system (ICS), or the incident management system (IMS). Firefighters are well versed in this management system, and many states require completion of these courses prior to earning the right to be an incident commander.

The training consists of classes numbered 100–400. Each block, in one hundred level increments, is progressively more expansive. The basic I-100 block is taught at many police and fire academies. Levels 200, 300, and 400 are progressively taken, and each level expands on the previous one. There are many components to this management system, and it is capable of accounting for thousands of responders working at one time. At large-scale incidents, law enforcement personnel will be prominently used, both as managers and responders. We need to get the supervisors involved in this type of training. These supervisors and responders will be part of the management team as well as front-line tactical workers. They need to understand the language, terminology, duties and responsibilities that are all components of this management system. A basic understanding will be required for the responders to work safely and efficiently in this system.

At all large-scale incidents, there is heavy police involvement, and, to work well as a team, the supervisors must be able to mesh with other responders who will be managing the incident. This is very important training, and the law enforcement responders need to be brought up to speed with others in the response community. Additionally, a key principle of incident management focuses on two areas where compliance of all responders will be critical. The first is called *unity of command*. The principle of unity of command holds that all responders working at the incident will answer to only one supervisor. Police sometimes may have more than one sergeant or other supervisor while on duty. However, while working at the incident, they will be given a specific assignment in a specific area. They will also be advised as to the one supervisor to whom they must report. In order for the IMS in place at the incident to work, all responders must remain focused and maintain strict compliance with the unity of command.

The next important principle of incident management is called *span of control*. This concept basically holds that supervisors cannot supervise too many responders, or subordinates, while maintaining proper control and accountability. Incident management allows a supervisor to maintain proper control of between three and seven subordinates. The preferred or optimal ratio between supervisors to subordinates is one supervisor to five responders. Often, on a daily basis, law enforcement supervisors will be responsible for more personnel than the IMS recommends. During an incident, there needs to be strict accountability of personnel who are working. In addition to unity of command, span of control is the other critical principle that must be followed. It is imperative that the ICS has the supervisors in place to assure all those working are properly supervised and that all of their needs are taken into account. Also, this strict level of supervision lends itself to better scene security, as all responders in an area will know who is to be there, and more importantly, who is not. This will be another level of security in attempting to eliminate or greatly reduce the possibility of secondary attacks or suicide attacks by additional terrorists. Further, this training must be instructor-led, and not taken online. We will discuss the differences between live, instructor-led training and online training classes later in this book.

Finding Time to Train

Now that we have identified several training classes and options for presenting the material, let's take a closer look at what is involved. Some states are "right to work" states, and others are "contract" states. Regardless of the type of employment rights, there will be certain issues when beginning to implement new training areas. In some instances, officers work an "unconventional" schedule, meaning either a three- or four-day work week that consists of anywhere from ten- to twelve-hour shifts. In these types of shifts, there are sometimes built-in "training days." These days or hours owed to the employer may vary from one eight-hour day per month to several times yearly. Should officers work these types of schedules; these training days can be utilized to present this new training to officers who are already working. If there are no such training days available, other options need to be explored.

There are many different ways to implement and achieve your new training mission. Unfortunately, for administrators, the best available training options may not be the most desirable. Training in these new areas can be done in conjunction with other training, which each agency is required to provide on an annual basis. For example, there are requirements governing firearms and other topics that are addressed at least annually. We can expand these training sessions to at least begin the process of integrating new training with our veteran officers. This solution will extend the training calendar in each agency, but again, we owe it to our responders to be as completely trained as possible.

Other training options include the dreaded overtime factor. While budgets are tight, we need to make training a priority for our police and law enforcement professionals. It is conceded that the use of overtime or "comp days" (a system where officers accrue hours to be taken as days off rather than to be paid an overtime wage) is the least desirable option, but please consider the alternative. Administrators are forced to battle with these financial issues each year. As budgets shrink and manpower numbers drop, administrators must find innovative and clever ways to meet demanding mandatory training requirements. Simply put, if we can find the time and resources for some training, we must find it for all training, regardless of whether it is new or old material, and regardless of its being mandated or optional. Agencies must begin to make this new and refresher training more of a priority. They need to heed past lessons, and not fall back into a reactive training state. Simply because we have not seen an active international terrorist attack since 9/11 does not mean that we have once and for all defeated terrorism. The groups still remain, their objectives remain unchanged, their resolve as strong as ever. History has shown a tremendous amount of patience on the part of terrorist groups. They have succeeded, to a large extent, on lulling us back into a false sense of security. Also, there still remains an often forgotten threat of terrorism from the many domestic groups that exist in America. Some of these domestic groups are extremely violent, or place emphasis on property attacks.

Should another international attack take place, what will be the response of our leaders? Americans will not tolerate another round of quotes such as "we didn't see this coming," "there were training shortfalls" or "there was a breakdown in our intelligence gathering." Americans will expect a far better and faster response. They have seen the press releases regarding the ridiculous amounts of money that have been allocated for homeland security.

As always, the first group of responders on the scene will probably be the local police or sheriff. They will be expected to take charge of the incident, call for resources and start lifesaving measures. We need to have them understand their responsibilities, duties and limitations fully. We cannot afford to have any more "blue canaries" become victims, thereby adding to our response crisis. These men and women need to be well versed in all areas of terrorism, and terrorism response. It is unfair to send them in their cruisers to an attack site and expect them to be able to properly assess the incident and then begin proper mitigation efforts. Based solely on their basic training and standard-issue uniforms, we can deduce that they are largely unprotected against any type of terror attack.

Should we be subjected to a CWA such as nerve gas, unprotected officers will quickly succumb to the agent used. Further, the initial signs and symptoms of exposure to agents such as sarin gas are very similar to exposure to pepper sprays. Untrained officers believing the sarin to be pepper spray may possibly lose not only their own lives, but the lives of countless others as well. Again, please understand that this is one chemical agent and one attack scenario in a sea of options available to the terrorist. We must present broad training across all CBRNE areas to best ensure safety of civilians and responders. Please, let us not forget that the police or first responders at the scene will be directing other responders into the area. A lack of understanding of attack methodologies and parameters could prove very hazardous, if not deadly.

We have explored but a few areas where law enforcement needs better baseline training. I am not an advocate of making all first responders robots, nor do I support trying to make all first responders interchangeable. Each response discipline must work very hard to maintain proficiency in their own areas. Forcing them to maintain multiple training certifications and licenses is difficult to achieve and not very realistic. However, as previously stated, there are gigantic holes in the training provided to our law enforcement personnel at present.

Medical Training for Police

In my experience, almost every state in America requires law enforcement personnel to maintain a minimum level of proficiency regarding first aid and cardio pulmonary resuscitation (CPR). In recent years, we have added the SAED to the cache of tools available to our law enforcement personnel. CPR techniques were dramatically revised not too long ago and are constantly being studied by the medical community. Certainly most law enforcement responders have been made aware of the new procedures and protocols for CPR.

Beat cops in all but the largest cities in America are most assuredly the primary provider of initial basic life support services. In smaller communities, and in communities where the medical responders are volunteers, the arrival time on-scene for these volunteers is far slower than in large cities with paid EMS responders. A police officer or deputy may be providing definitive, necessary medical care for many tense minutes while awaiting resources from a volunteer first aid squad. In my region, fire and first aid services are almost exclusively provided by volunteers. Given the current state of affairs in America, we have seen a significant decline in the number of volunteers.

Years ago, there was a single breadwinner in most families, and the spouse of the breadwinner was able to maintain the domestic duties of their respective households. With financial obligations increasing, stay-at-home parents were no longer possible for many families. The parents were both forced to turn to outside employment to either meet the bills, or to provide a small financial cushion. This turn in familial dynamics has had a tremendous negative effect on volunteerism. To make matters worse, in many families at least one of the parents now works a second job. The precious free time adults have for family obligations are not about to be given away for free. Time requirements to maintain membership and training levels for fire and first aid positions have grown burdensome. Accordingly, there are fewer responders.

How does this affect cross-training police? Simply put, the response time of certain resources is staggeringly poor. In fact, in some places, daytime calls for service sometimes are not met, resulting in some jurisdictions being forced to hire responders to replace the shrinking volunteer rolls. Additionally, most people do not work in the neighborhoods where they live. As volunteers, they are unable to leave work to answer calls for service. There is virtually no financial reward for volunteering one's time and in fact, volunteerism has the reverse effect in that it takes time that could otherwise be spent earning money. Therefore, we need to ensure that street cops and deputies have a firm background in first aid protocols and techniques. They may be waiting much longer for assistance to arrive than their predecessors did in years gone by. But have we provided the officers with additional training to bridge the gap? More likely than not, we have failed in this area. When someone calls 911 for a sickness or injury, they anticipate a qualified, dedicated and rapid response to their call. Citizens will certainly call for our police to be on-scene and treatment ready. This adds to their required annual training. How do agencies meet this need? As already stated, administrators need to make it a priority. But how many priorities can an agency have? Taxpayers want the service, but are reluctant to pay for it to any degree over what they already do. This creates a conundrum for agency heads, as they are asked to provide faster and more comprehensively trained responders, but have their financial hands tied.

Learning from the Past: Response to Explosive Incidents

We have discussed some possible areas for cross-training police where one can envision the benefits on a daily basis. There are other areas where we need better training that hopefully will never be needed. Large-scale terrorist attacks have rarely occurred in America. Timothy McVeigh unleashed a deadly attack in Oklahoma City in 1995, and obviously, there was 9/11. While the Oklahoma City incident was severe, it was not on the same scale regarding loss of life as the events of 9/11. On September 11, 2001, the response communities were caught off guard. Should we experience another incident, we will be expected to do much better. The only way we can respond better is to be better prepared. The only way to be better prepared is to train. But we do not have the time and we do not have the resources.

This scenario is totally unacceptable for our responders and our citizens. How can this situation be resolved? Simply put, there is a need to make training a priority. As the saying goes, you never know what you have until it's gone. We do know what we have regarding training, and we need to make it better. Imagine the public outcry if we are attacked again, and the response is no better than it was on 9/11. What will be used as an excuse? We have had *only* eight-plus years to correct the problems that were exposed on 9/11. The fact that significant issues remain after all this time is totally unacceptable. We have been fortunate to have identified the problems. We need rank-and-file law enforcement professionals to be more attuned to their surroundings and able to protect and serve our communities.

Regarding the threat of international extremist terrorism, officers need to understand the extremists' culture, so as to not offend any members of the community and make intelligence-gathering better. Is this training provided to all law enforcement professionals, and is it consistent among agencies? Maybe, but maybe not. Are our police aware of the preferred training methods of terrorists, and are they in touch with potential terrorist indicators? Certain chemicals when mixed together are the basis for explosive solutions. We must make our responders aware of the common ingredients, and the chemicals used to make explosives. In one of my training classes, we were taught that urine can be stored (as a replacement for urea) as a component to make a bomb. It can be a substitute component in urea nitrate, which was used in the 1993 World Trade Center bombing. After Oklahoma City, officers were looking for large rented trucks as a bomb delivery system. Other training suggests that the same 4,800 pounds of ammonium nitrate can be hidden in a full-size pickup truck, or in a limousine. Has anyone advised our law enforcement professionals of this fact? Are we providing inaccurate or dated information to our police responders? Who is responsible for providing accurate and up-to-date information? Sadly, there is no consensus as to who is responsible.

Continuing on the theme of explosives, one can delve deeper into another area where we are lacking in accurate information and in comprehensive training. Many times beat cops are sent to suspicious package calls. The officers arrive to find some sort of package or container that has been discarded or abandoned. They will generally take a close look at the article, often walking right into the "kill zone." But why does this happen? Have we not trained our officers to remain at a safe distance, and then to call for the proper resource to "clear" the package? Most civilians, as well as some deputies and officers, are not even aware of the correct asset to call to the scene. The consensus of people asked will generally respond that if there is any doubt as to whether a package is a bomb, the bomb squad or a bomb technician should be summoned. However, this opinion would be inaccurate. The correct responder to request would be a K-9, specifically a "bomb dog." The canine units, consisting of the dog and a handler, are trained in explosive detection. Unlike on television, "explosive detection K-9s" will not claw at a package determined to have explosive components. Rather, the dog will sit down to alert the handler to the presence of explosives. Accordingly, once a K-9 unit determines the possibility of explosive elements, a bomb disposal unit will be called. They will conduct further examination and investigation of the article in question. These techniques may include x-ray examination of the package.

For example, if there is a reported bomb threat in a building, certain steps need to be followed. Many times, law enforcement will be the first to arrive at these incidents as well, and be expected to take immediate decisive action. And what will they do upon arrival? If the building is empty, one possibility is to conduct a cursory search of the building. But what are those delegated to search the building looking for? Today, bombs are constructed in a fashion very different from the public perception. The days of a spherical metal device with a smoking fuse are over. Bombs and bomb makers in our technological present are very sophisticated. Many typically found items in a building can be improvised so as to be used as explosive devices. Are the officers assigned to the search team trained to not use their radios? Most probably are, but at one of my training classes, we were directed to leave them outside the search area, because, even if they are off, they can detonate the bomb if one is present. Are law enforcement personnel aware that static electricity, such as generated by dragging our feet on carpet can act as a stimulus for explosive device detonation? Again, some may be aware, while others may not. All must be aware, and supervisors need to be confident that subordinates are so trained.

The most common type of bomb found in these cases is generally referred to as a "pipe bomb." These devices are made of polyvinyl chloride (PVC), aluminum or steel pipes filled with an explosive and shrapnel, such as glass, nails, ball bearings, etc. The stronger the casing, or the pipe, such as steel, the more devastating the results will be. But we also know that at certain times, criminal elements are aware that we will uncover their bases, where they conduct their criminal acts. So will they leave us surprises? You can be certain that some will. A flashlight is an example of a common piece of equipment that can double as a pipe bomb. The components are there, a cylinder, hollow in the middle, with end caps. Add explosives and shrapnel, and, voila, the flashlight is now a pipe bomb. No "good" cop can walk by a flashlight without picking it up. If it is equipped with a mercury-type switch, the mere movement will cause detonation. If that is not the detonation sequence, here is another way. After the deputy or officer picks up the flashlight, the next step, naturally, is to see if it still works. Press the button to turn on the light, and there is your detonation sequence fulfilled. And of course, the officer is looking into the bulb, to be sure the light works. How many police officers are aware that criminals sometimes leave "pipe bomb" flashlights for us to find?

There are a great many other examples for us to consider in this area. Some bombs are set to detonate by photocell. With this type of explosive device, if you block the sun or light source, the bomb explodes. Letter bombs, book or magazine bombs are also a possibility. Certainly, there are certain types of magazines that officers may wish to "investigate." These can be converted into bombs using readily available explosives. Law enforcement personnel receive very little explosive training in basic police academies. This is another area where poor training can prove fatal for our responders and citizens. Earlier, I referred to a call to an unattended package. In a previous training class, I was shown a video. The scenario was in Canada, and the officer was on his way to the station on his last shift prior to his leaving for vacation. He reluctantly responded to the call of the abandoned package. He got out of his cruiser, and went toward a small bag, similar to a brown sandwich bag. He picked it up to investigate, and it detonated. The reluctant officer with his mind on vacation was tragically killed. I wonder if he was properly trained and ignored his training, or if he was untrained and simply in a hurry. Sadly, we will never know. There are hundreds of other examples, but the message will always be the same. We have not provided all the requisite skills for our officers to survive their shifts.

Improvised Explosive Devices

There is a relatively new trend in the world of explosives and explosive attacks used against us. The term and concept is improvised explosive device (IED). We have read accounts of soldiers in Afghanistan and Iraq who have been killed by "roadside bombs," or IEDs. This trend is just as likely to kill law enforcement responders at home as it is to kill our brave solders serving overseas. By its definition, any device or article improvised or altered to house explosives is an IED. As already stated with the flashlight, magazines and other items, IEDs are easy to make, and very deadly. During a building search, almost anything can be considered an IED, provided we find explosives in it. Another new term is vehicle-borne improvised explosive device (VBIED), or large vehicle-borne improvised explosive device (LVBIED). Timothy McVeigh's rented truck was a VBIED. A car bomb is a VBIED. In Columbine Colorado, the students left propane cylinders in the parking lot to serve as VBIED. Fortunately, these devices were not detonated. But again, these are but

Figure 2.6 Thermal imaging cameras are commonly used by firefighters to locate victims during firefighting operations. However, they are a great tool for law enforcement use as well. They can be used when attempting to locate lost persons and can be used for tracking suspects at night. They work by detecting heat, and the hotter an object is, the brighter the image will appear. Some cameras are available in color, have a digital temperature display, and can transmit images to a receiver at a remote location a short distance from the camera.

a few real-world examples of IED implementation designed to kill. Car bombs are prevalent in the Middle East, and there is always a chance that these proven terror tactics will be brought to bear on the United States. Will this be another area where America's responders are reactive? Will responders and citizens perish prior to implementing training for police? Only time will tell.

Surveillance and Counter-Surveillance

Continuing on the same path of needed training, we need to propose counter-surveillance training for our police, so that they can be aware of when someone is watching them. Inside the Al Qaeda Training manual, there are entries that instruct terrorists on surveillance and counter-surveillance. Amazingly, the terrorists are training to watch us, and make sure that we do not know that we are being watched. Are we taking these same training steps? Not all officers are. Where will the next terrorist attack be? I wish I knew, so we could stop it and prosecute those involved. But none of us is aware of when or where it will come. We know Al Qaeda has invested time in surveillance of small towns, as well as targets in large cities. So we must *all* be ready for the next attack. We also cannot forget the previously mentioned threat of domestic terrorism. We saw domestic terrorism in Atlanta at the Olympic Games, where a backpack device was detonated. Also, Eric Rudolph was sentenced to life when a police officer was killed responding to an abortion clinic where Rudolph had planted a bomb.[1] Our adversaries take the time to conduct surveillance training, and we sit back and hope nothing will happen. Who is proactive and who is reactive?

Chemical Agent Attack Training

Another area that we are seemingly deficient in would be in potential indicators of certain types of attacks. For example, one of the CBRNE attack possibilities centers on chemical attacks. These potential attacks can take many different forms. CWAs, which include nerve agents such as sarin, soman, tabun, and VX, may be used against us in confined spaces. All of these agents are synthetic, and would be aerosolized for deployment against us. One of these agents, sarin, was used in Tokyo, Japan, in March 1995, in several subway lines. There were eight deaths, and thousands of people who sought medical treatment.[2]

Another type of chemical attack involves the use of vesicants, or blister agents, which are also aerosolized for deployment. The main vesicants available for terrorist uses are mustard gas and lewisite. Each of these agents is very lethal in its own way. There are telltale-warning signs for each agent as well, such as odors, which are produced by some of the agents. To protect not only our responders, but civilian populations as well, we need to train our responders about these indicators and odors. While we may have been given some form of training, it needs to be comprehensive and refreshed as needed. Training our responders on exposure to indicators once in a twenty-years or longer career is not going to be adequate. Signs and symptoms of exposure to these CWAs also need to be taught to our personnel, as well as arranging for refreshing the training. For example, law enforcement personnel are often required to carry some form of incapacitating chemical agent, such as mace. More likely, today's officers will carry a variation of oleo capsicum (OC) spray in canisters that contain either a liquid or foam form of a solution derived from pepper plants. When needed, the canister is taken from its holder, and the contents are sprayed in the face of the unruly individual, generally just above the eyes. The agent produces certain specific immediate (usually) reactions from the individual who was sprayed. These symptoms include the eyes' "slamming" shut due to a burning sensation; a cough, or tickling sensation in the back of the throat; tearing of the eyes and a discharge from the nose. These symptoms generally recede after a short time, and the targeted person can have these symptoms eased via decontamination with deionized water applied by either police or first aid responders.

What is of note here is that initial exposure to the nerve agents explored before are strikingly similar to those of OC spray. The only differences occur after immediate exposure. With the CWAs noted, the next level of symptoms after exposure is likely to be very severe or lethal if an adequate dose is absorbed. The next level includes seizures, vomiting, respiratory arrest, and possibly death. There is quite a variation between exposure to OC sprays and CWA exposure. But if law enforcement responds to a location believing that a release of OC occurred, but instead a CWA was released, the consequences may very well be fatal.

In addition to the lethal chemical agents designed to kill us, we have to factor in when chemicals are used to be weapons of mass destruction (WMD). Ask police or sheriff's deputies what a toxic industrial chemical (TIC) or a toxic industrial material (TIM) is, and you may get a blank stare. Basically, these are chemicals and materials that can produce death or serious injury after exposure. In Iraq, we have seen chlorine trucks modified to become IEDs.[3] Fortunately, the terrorists have not determined the proper amount of explosive to attach to the chlorine container to effect a compromised container, resulting in the release of chlorine. The explosive charges used to date have consumed the chlorine after detonation, thus not releasing enough chlorine to produce fatalities, as desired. While chlorine has a smell that most Americans can easily identify from swimming pools, the other chemicals and agents may not be so easily recognized. CWAs, TICs and TIMs are all readily available to the terrorists, and we must prepare our responders to recognize them, lest they conduct tactical operations wherein tragic and fatal decisions are made.

Biological Attack Training

We have explored many different areas where training of law enforcement officers is deficient or completely lacking. Still other areas can be explored. The first is one where there is not much direct action that the law enforcement community can act upon. If we are subject to a biological attack, the initial responses would come from Public Health and the first aid community. These biological agents, if used, would probably take days to make themselves known to our medical professionals. These agents will require some time to incubate before we realize that they have been used against us, and that we are required to take action. The government has set contingency plans for such a nationwide or even regional outbreak. The plan would involve determining the type of agent involved, and a medical countermeasure or prophylaxis would be developed and distributed. Should the public grow impatient while waiting for the prophylaxis to be manufactured and distributed, civil unrest would probably follow. We would require significant numbers of law enforcement resources to maintain peace and calm during this time.

Further, the law enforcement community would be called upon to secure and protect the medication supply prior to and during its dissemination. Short of this type of service, it is unlikely that law enforcement would have significant responsibility during a biological attack. We should, however, train our law enforcement

Table 2.2   Sampling of Potential Biological Agents

Agent	Type	Contagious	Mortality Rate
Anthrax	Bacteria	No	80%
Plague	Bacteria	Yes	90%
Small Pox	Virus	Yes	30%
Ebola	Virus	Yes	90%
Botulinum	Toxin	No	80%
Ricin	Toxin	No	High

responders about the potential for biological attacks. Certain biological agents, such as ricin, a toxin derived from castor beans, have been processed in America already. In the United States, we have seen ricin produced for use against law enforcement officers. Ricin is a very deadly toxin, requiring minute amounts to be ingested to produce death. There is no known antidote for this toxin. It is manufactured to resemble salt or sugar, and can be easily secreted by someone with evil intentions. It would be logical for law enforcement officers to be frightened during a biological outbreak. With training, they can be assured that with the required prophylaxis, they can be safe and protect their families if they come to work. Without this training, an increased number of officers may call in sick.

Last, while most biological agents require prolonged incubation time, some do not. Additionally, while some are reasonably fatal, others are used as "nuisance" agents, designed to make the intended victims ill, but not to kill them. This was evident in Oregon in 1984. During this event, followers of Bhagwan Shree Rajneeshee delivered salmonella via spray bottles to ten buffet restaurants. The salmonella was sprayed on salad bars, and affected 750 individuals. The purpose of this attack was to influence a local election, hoping that Rajneeshee's followers would win two of three contested spots for circuit judge. No one perished, but this was an example of a biological agent used as a nuisance agent.[4] Table 2.2 shows a small sampling of the many possible biological agents of varying types, contagiousness and mortality. A national pandemic or epidemic of any of these agents could result in thousands of deaths nationwide. Most states have plans for dealing with the potential outbreak of these agents.

Radiological/Nuclear Training

The last of the CBRNE scenarios we will discuss involves radiological, and then nuclear incidents. There are several scenarios where radiological materials can be used against us. The most common would be a "dirty bomb" scenario. These devices are technically known as radiological dispersal devices (RDD). Many in the general public mistakenly believe that RDDs are the same as nuclear devices. This, however, is not the case. RDDs are radiological materials surrounded by conventional explosives. These can be set up in a number of ways, including the previously mentioned pipe bombs. Radiological materials are present and readily available from a number of sources. Smoke detectors contain small amounts of material, as do hospitals. The methodology to construct and use an RDD would include taking the radiological materials, loading that into the bomb casing along with the explosive, and detonate. The purpose of an RDD is not to kill great numbers of people. Rather, based on the size of the device and amount of explosives used, only those people in proximity to the device at the point of detonation may be killed. The primary reason to use an RDD is for "area deniability." This means that we would have a contaminated area that we would need to spend time to decontaminate, and, based on the readings taken, we might need to evacuate the area.

Many law officers have heard the term, but do not even know what an RDD is. They too share the misconception that it is a nuclear device. In fact, very few law enforcement officers have a firm understanding of radiation at all. There are four types of ionizing radiation. These are alpha, beta, gamma/x-ray and neutron. Alpha, beta and neutron are particles, and gamma/x-rays are waves. Radiation is measured by dose rate, which is the speed at which the radiation is coming at someone, and in total dose, or the amount of radiation one is exposed to. There are various manners in which we can detect this radiation. The easiest method would be to outfit the officers with radiological detection pagers. These devices are similar in size to the personal pagers that many people wear today. They have various options, but many can detect the rate and dose to which the wearer is exposed. There are training courses available to provide a sound understanding of the principles of radiation, and these same classes can offer education on selection, use, calibration and maintenance of these pagers. Training classes, courses and options will be discussed in great length later in this book.

There are other detection tools, but they are larger, more complex and require additional training. Additionally, during certain training classes, the students will be exposed to decontamination techniques. Responders should be familiar with federal standards for a responder's lifetime exposure to radiation. Agencies should review the federal guidelines and establish policies regarding the exposure of their personnel to radiation, then establish firm career exposure limits. Most administrators are unaware of these policies, and most of their personnel are completely in the dark regarding them as well. Should officers respond to an incident and be told that the rate of radiation is 30 micro-rem per hour, would they know what that level means to them? Can they deduce if this is a high rate, or a low rate? Does this piece of information require immediate action by responders? The answer to these questions is that it is a very low rate, just above certain levels of radiation always present called "background radiation" similar to the low levels of radiation present in our environment from natural sources, such as radon and the sun. But most officers will not know this. This information is very important and needs to be presented to our responders.

The last area within the CBRNE realm is nuclear attack. Due to the dynamics of a nuclear attack, there is not much anyone will be able to do in the hours after the blast, or release. All we can reasonably expect from any of our response disciplines would be for them to stay out of the hot zone and to assist survivors in getting definitive medical care. Short of that, not much else is practical for anyone to attempt. The good news for us is that our nuclear arsenals worldwide have redundant safety features built into them. This fact makes a nuclear attack by means of a nuclear facility on our soil very unlikely, though, I must concede, it is possible. A number of "suitcase" or tactical nuclear weapons are rumored to be missing from the former Soviet Union. Should a terrorist group, foreign or domestic, be able to obtain one of these weapons, a detonation would have very serious repercussions. These small nuclear devices are very powerful, and would inflict great damage in terms of loss of life as well as destruction and contamination if it were detonated in a crowded environment. Again, responders would be limited in their response strategies and tactics. While some may claim to have "radiation suits," there are no real garments that would protect the responder from the effects of exposure to radiation after a tactical nuclear device detonation.

Special Operations Training

There is one other specialty area to which law enforcement needs to have access. In some areas, as we have already discussed, SWAT operators respond to barricaded subjects, hostage crises and related incidents. These desperate criminals and disturbed people will sometimes start fires while the incident is ongoing. Further, the potential exists for the suspect to await entry of SWAT personnel prior to igniting the fire. If SWAT teams enter a residence that has been properly prepared by a deranged subject, they may be unprepared for a speedy exit if an emergency arises or conditions critically degrade. If SWAT officers enter the residence and the suspect has left accelerants on all floors, a fire, if ignited would spread very quickly. Even if SWAT personnel were equipped with, and wearing APR, they would probably succumb to the deadly by-products of the fire. Their APR would not provide adequate respiratory protection from the fire. We have seen cases where drug dealers "booby trap" their structures used for drug production, so it would not be unreasonable for a fire scenario to unfold as well. Some SCBA manufacturers make small, lightweight products for SWAT-type law enforcement response, but not all teams realize the importance of breathing apparatus for their teams. Further, teams would need to practice shooting their weapons while wearing the breathing apparatus. The mask or face pieces required when wearing SCBA will have an impact on the ability of the wearer to shoot in a manner consistent with regular firearms training. These specially trained SWAT operators should get some basic firefighting training, such as how to exit a residence under fire conditions. Night vision equipment will be of no use in a fire, so if the retreat plan includes this tool, it will fail. Some basic fire training in construction, search and rescue and preplanning structures could prove life saving.

Training for Corrections Officers

Another group of professionals in need of cross-training is corrections officers. These professionals are employed at the county, state and federal level. These trained responders make up a large pool of potential responders in case of serious or large-scale incidents. Laws will vary from state to state, but some of these officers are permitted to carry firearms either on or off duty. Many receive training outside the normal scope of daily operations that occur inside jails and prisons. While there will be logistical issues related to using these officers to supplement law enforcement's response, providing them with training in all of the aforementioned areas may reap great rewards in case of an incident. Corrections facilities may be placed in "lockdown" to free up many officers to assist law enforcement officers in the field. Another logistical issue would involve transportation and communications with outside agencies. These issues can be easily addressed if administrators are made aware that their staff may be called upon to assist in certain cases. Plans and arrangements could then be developed and implemented. It is up to our leaders to recognize all assets available to them, and properly prepare and train their personnel to respond when called upon to do so.

Additional Training Needed for 911 Operators

Finally, there is one more group of people who should be exposed to different training courses, and it seems logical to present their case under law enforcement. These groups of professionals are referred to by many different titles, from dispatcher, to 911 operator to public safety telecommunicator (PST). For responders on the incident, the PST is their lifeline. These are the people who will forward the initial reports and observations made to additional incoming units. They will dispatch other responders from different response disciplines to the scene. They will need to understand the information presented to them prior to forwarding it to the proper incoming responders. They have a tremendous responsibility, and this community of professionals should not be overlooked when planning and scheduling training classes. At a minimum, they should all be trained in hazardous materials and receive annual refresher training. Certainly CBRNE training would also be beneficial, as the more trained professionals monitoring signs and symptoms of victims, the better. The PST will also be charged with ordering the requested resources to the scene, until a formal logistics section under the incident command system is developed. Once it is developed, the responsibility for ordering resources may be transferred to those on-scene. To exclude these PSTs from training would be a mistake.

Summary

In this chapter, we have explored many areas where law enforcement professionals and others in the general field of law enforcement are in need of initial or refresher training. We have spoken about hazardous materials calls, fires, CBRNE, IED scenarios, and incident management. We have delved into carbon monoxide responses and SWAT presence at houses potentially "prepared" for arson. Since we need to be prepared, we need to refresh all existing training, and we need to delve into all other relevant areas. Administrators need to take a good, hard and impartial look at the current state of our training. If they do as they are requested, they will see that we are not nearly as ready as we should be. They willingly send their officers out into the world believing that they are able to respond to and handle anything that may come up. In reviewing the materials presented here, I believe that we can quickly and easily dispel that notion. My goal is to have enough decision makers see the shortfalls that have been identified and presented, and make genuine attempts to remedy them. It is conceded that money is short, manpower is reduced and time is simply not available. These are exactly the reasons we need to expand our training capabilities. When we get the unthinkable assault again, we will probably be at reduced staffing levels, still in need of money and still wishing we had more hours in the day. But we won't have those needed hours. There will be even fewer of us on the streets trying to deal with casualties, panic, mayhem and repeated requests for assistance. We owe it to those who will be in the field that day to give them all the tools they will need to draw upon for a successful resolution to the problem or incident presented to them. As of now, I believe that most of the responders will be coming to the scene at code-three speed, but with an empty toolbox. That alone should prompt us to act. Anything else will be unacceptable.

Endnotes

1. "Rudolph gets life for Birmingham clinic attack," CNN.com Law Center, Monday July 18, 2005. www.cnn.com/2005/LAW/07/18/rudolph.sentencing.
2. Lawrence K. Altman, "The poison: Nerve gas that felled Tokyo subway riders said to be one of most lethal known." *NY Times* World, March 21, 1995. http://www.nytimes.com/1995/03/21/world/terror-tokyo-poison-nerve-gas-that-felled-tokyo-subway-riders-said-be-one-most.html?scp=9&sq=Terror+in+Tokyo&st=nyt.
3. Damien Cave and Ahmad Fadam, "Iraq Insurgents employ chlorine in bomb attacks" *New York Times* Middle East February 22, 2007. www.nytimes.com/2007/02/22/world/middleeast/22iraq.html.
4. John Mintz, "Technical hurdles separate terrorists from biowarfare." *The Washington Post*, December 30, 2004. www.washingtonpost.com/ac2/wp-dyn/A35011-2004Dec29?language=printer.

Chapter 3

Can't We All Just
Get Along?

In Chapter 2 of this book, deficiencies in the training of our law enforcement responders were explored. These dedicated men and women do a very professional job under extremely trying and stressful conditions. There are various ways some professional responders cope with the stresses of the job. Some of these are not very healthy, such as drinking, smoking and gambling, just to name a few. Others choose to deal with these issues by turning to humor, often referred to as "gallows humor." The events seen—the injuries, the death, the pain, the constant reminder of just how cruel humans can be, and how tragic life can sometimes be all lead some responders to have a cynical and sarcastic outlook. These people throw out comments in the face of these events that would astound most non-responders.

What Makes First Responders Tick?

Most of the responders in America, as already mentioned, are of the Type A personality. They are people who are aggressive, spontaneous, curious, authoritative and demanding. Sometimes, they use gallows humor in the presence of others in the response community. As a rule, responders from each discipline will privately acknowledge the need for each other, and even admit to respecting the job done by each other. There are those who have problems dealing with blood and prefer to not get involved in certain first aid matters. Therefore, one can truly appreciate those who arrive at a serious motor vehicle accident and begin treating lacerations, broken bones and other assorted serious or horrific injuries. Others are fine with traffic control, preparing the vehicle for removal, fire suppression, vapor suppression or even the mundane task of applying the absorbing agent to collect released fluids. Some have commented that they do not know how EMS workers are able to do that job. Conversely, many in the EMS community will respond that they do not know how firefighters run into a burning building while everyone else is running out. A firefighter's reliance on technology, tools, training, and equipment makes this task one that is easily embraced. This is similar to the way police respond to a shooting or an officer-down call with shots fired. Everyone else is ducking for cover, yet the police officers are racing toward the danger. Virtually no one wants to be present for the hazardous materials call, where there is a large release of what firefighters call "hazmat." Police will anxiously await the arrival of the hazardous materials team and rely on their expertise in mitigating these incidents. The hazmat teams have even cooler stuff than the fire department.

Why Are There Hostilities?

The various response communities can all agree privately that they need each other. Why then are there so many issues among the agencies and even the responders themselves? It can be agreed that there are certain issues related to the gallows humor, but sometimes there is just genuine hostility among certain members of the response community. For example, in my case, there are those in the law enforcement community who tell me to "make up my mind" regarding my volunteer activities. If my fire engine were to arrive at a house fire and I offered the police officer on scene my turnout gear, he would politely (I hope) refuse. Similarly, I would not don a first aid jumpsuit to treat a seriously injured person if the first aid squad were there. Each individual is aware of the limitations of their training and their desire to remain working in their own area of expertise. There are many other examples where different response communities interact. Yet there remains some hostility among members of different response disciplines. The question then is, why can't we all just get along? Recently, at a training class attended by both law enforcement and fire personnel, one of the police officers referred to the fire personnel as "hose dummies." He went on to explain how dumb he felt the firefighters were. Fire personnel were quick to reply. They claimed that police do more harm at fire scenes than fire personnel do at police scenes. The cop disagreed, but the firefighter responded by saying that fire crews are not dispatched to bank robberies. There seemed to be general hostility between this officer and the firefighters. Sadly, this type of scenario is played out all too frequently.

Another perfect example of this phenomenon was seen in New York City when the focus of the world was upon it during and after the tragedies of 9/11. In the months after that incident, a review of emergency services was undertaken in that city. The New York City Police Department (NYPD) has a division of its force called the Emergency Services Unit (ESU). The members of this team are very well trained, skillful and professional. They represent my premise of cross-training very well. They have access to weapons systems that are superior to the rank-and-file's, are trained in SCBA, high angle rescue and other specific areas that garden-variety patrol officers are not. Similarly, the Fire Department, City of New York (FDNY) has a core unit in each of the five boroughs that has undergone additional extensive training. They are the five FDNY rescue companies. They are trained in many of the same areas as the ESU members. In fact, ESU and FD rescue personnel have been simultaneously dispatched and have responded to the same incident. Upon arrival, there was sometimes debate as to who was to be in charge of the incident. Further, there were disputes regarding which service was going to handle the tactical or technical aspect of the incident. Between these two highly trained specialty units, tensions were high. Both units had similar training and experience. Both squads of responders were adequately trained and able to handle the assignment. Yet, they could not agree among themselves who would handle each type of call. Often, the group of responders that was first to arrive would begin to mitigate the incident. Certain calls, such as hostage situations, barricaded subjects or fires obviously required one of the teams to prevail. But for a long while, genuine dislike for each other existed. Even though each of the teams had the same goals of safe, professional and speedy resolution of the crisis, an agreement could not be made regarding which agency would have incident management and control. Fortunately, there were never situations where this adversarial relationship affected the outcome of a call for service. This breakdown in the response to these emergencies resulted in a review of the response protocols by city officials, and the issue has now been resolved.

Further inspection of the NYPD and FDNY will reveal other issues between the two agencies. Every branch of emergency service has certain traditions that they hold somewhat sacred. Very often, the response disciplines are unaware of these customs and traditions, and this may be a cause for tensions between agencies. These traditions hold core values of the responders, and to ignore or dismiss them is an insult of the highest order. One of the most revered customs or traditions in the fire service, for example, is the recovery of the remains of a fallen firefighter. Tradition holds that once the remains are located, the surviving crew members, if able, are to secure and remove the remains. In career departments, firefighters are assigned to platoons, tours or shifts, and generally work out of the same fire station. If the surviving crew members are unable to remove their fallen brother or sister, members from another tour in the same station will do the "honors." At ground zero in New York City, a somber, spontaneous "remembrance" was observed on the "pile" each time the remains of a victim were discovered. The remains were placed in a Stokes basket, or litter, all work on the pile would stop, and the remains were removed to street level and transported. A lack of understanding of the depth of the firefighters' custom of victim recovery reportedly led to several altercations between fire service personnel and other workers there. Other workers on the pile, including NYPD members, did not want to halt work on the pile after the remains of a firefighter were discovered. Once the body was identified, possibly by accountability tag, or by the front piece

Figure 3.1 This symbol, not always located near doorways, will advise fire-fighters that the structure is unoccupied. It might be overlooked by untrained responders who may not be looking for such a marking. It is generally used on abandoned structures that have been searched and cleared by fire personnel. All responders need to be aware of the marking and its meaning. Do firefighters just assume everyone will be looking for these symbols or markings, or do they seek out other responders and alert them to their meaning and locations?

worn on the fire helmet, members from that company (a name given to a fire crew, based on their designation) were notified, so that they could "bring their brother home." The passion of the firefighters to follow this tradition was met with the need of other responders to continue their grim task on the pile. Occasionally, the remains of firefighters were discovered and removed by workers who were not from the fire service community. This act would spark outrage among the firefighters. Based on these examples, we can see how we sometimes simply do not get along, and sometimes do not fully understand our brother and sister responders' values and traditions. While these examples were in New York City, they are not being singled out. The mere fact that those two departments are so large and located in the media capital of the world, means that they are always under scrutiny. Those responders have my utmost respect and admiration and do an excellent job.

In my home state of New Jersey, some emergency tasks are shared among responders as well. For example, motor vehicle extrication is handled by some fire departments, and in other areas, that task is left to the first aid squad. In fact, some of the area squads are called "rescue squad," not "first aid squad." Here again, it can be seen how this can be a recipe for disaster. Responders who are trained in an area, but do not perform those tasks are quick to cast a watchful and critical eye over those conducting the operation. No one likes to be second-guessed, or be subject to "Monday-morning quarterbacking." However, this will sometimes happen, and when it does, there are usually hurt feelings. Worse yet is when agencies are conducting an operation and the "back seat driver" is behind you with suggestions that serve only to distract you from the task at hand. Those who respond to and observe emergency service calls are often referred to as "buffs." Again, this type of behavior by these buffs lends itself to animosity and bad blood between agencies and departments.

In general, there is nothing but respect for those in our military service organizations. Responders can certainly empathize with the time away from loved ones, difficult duty assignments and the constant danger they face. While there is some banter among the different branches of military service, there is none of a serious nature from the first-response community. Additionally, the same can be said of the hazardous materials response community. There is some light teasing from police and firefighters, but in general, hazmat personnel receive a good amount of respect from those responders. When the hazmat technicians are needed, it is all business for them, and they are certainly appreciated. There will always be those who will tease others in the field, but sometimes this teasing gets out of hand. In certain areas, the ribbing gets malicious, and in fact, there is often genuine bad blood among some responders, either within the same discipline, or among disciplines. In my experience, relations between police and fire are better in small towns than in the larger cities, though I do not know why.

Intra-Discipline Issues

There are also times when we do not get along with those in the same response discipline. It has already been explained the personality of most responders is Type A. These types of people are very competitive. There are cases where members of one fire department or first aid squad simply do not like members of a nearby or neighboring agency performing the same job. They will be quick to talk badly about the members of the other agency. They also offer quick and biting criticism of the "rival" agency. Tactics, strategies and final results are all criticized in an effort to promote themselves or their agency, and to discredit the other agency. For example, a harsh criticism of a fire department would be that they are only good for "saving foundations." This would imply that their tactics allow the fire to get away from them, resulting in a total loss of the structure except for the foundation. Advise firefighters that their company has that reputation of saving only foundations, and watch the anger and disbelief in their faces. Sometimes, individual firefighters will be accused of "being afraid of fire," or of being "afraid to go in." Again, these are very serious claims, whether they are true or not, and will be met with strong denials or confused looks. The same can be said of policemen who would not like to be told that they "have no heart," or that they are "punks." This would mean that the officers are not likely to provide speedy backup when another officer is under physical assault. Again, is this competitive banter, or is there genuine dislike among members of the same service? Often, the remarks are out of jealousy (their agency may have a larger budget for equipment, or they may be better paid, or get more work), and sometimes they are part of a prevailing culture of animosity. I have heard of neighboring fire departments who for one reason or another will try to avoid calling each other for assistance, even though they are close by. Fresh people and ideas have mostly resolved these long-standing sad situations. When I have inquired why these things happen, I have been told that "I don't know why, we just don't like each other." These issues need to be quickly identified and resolved among the responders. Perhaps there was a misunderstanding between the agencies years ago that was left unresolved. The bad feelings may have lingered, and the relationship deteriorated beyond repair. We need to repair these damaged relationships.

Volunteer Professionals versus Career Professionals

Another area in which we cannot get along is between career or volunteer, full-time or auxiliary responders. This culture of dislike is most prevalent in the fire service, but some exists in law enforcement. In many states, if not all, there are fire unions and organizations. These organizations sometimes seek to promote the career firefighters, at the expense of volunteer fire departments. Generally, most fire agencies are constituted in one of three ways: (1) there are fully paid career departments, as in most major cities; (2) there are volunteer-only departments, which make up the majority of the fire service, and (3) there are combination departments that offer a mixture of paid and volunteer members. In the fire service, there are many issues between the career members and the volunteer members. While this may not be the case everywhere, there are certainly many places where this is evident.

In some areas, you must have served on the volunteer fire department prior to be considered for a career position. Once a career position opens, the department or district will follow their hiring protocols. These protocols come in many fashions, including "civil service" exams; "chief's tests," interview, appointment, etc. There are generally more than a few qualified and eager applicants for the few open positions. Once volunteers become paid members, they may be required to become affiliated with one of the labor unions or fire service organizations. This is generally done for purposes of union and contract negotiations, for protection from disciplinary or legal actions or for job-related incidents.

One disturbing trend I have seen is that once a volunteer gets his paid position, the remaining volunteers who are left behind are often referred to as "scabs." This is a very offensive name to be given to someone who volunteers time and effort while doing a very dangerous job where there may be no career members available. This is not the case in every agency, and there are a great many agencies that exhibit tremendous respect between volunteers and career members. What a great many of these career firefighters forget is that many of their members were volunteers at some point prior to their career appointment. Some volunteer members are trying to "get a foot in the door" in order to be hired with that agency in the future. The fact that they are a volunteer member today and career member tomorrow seems lost on some of the people with a dislike of volunteers who are commencing their employment.

Many volunteers, and I know many, who are awaiting a chance to become a career firefighter, are abused and called names while they are waiting for their opportunity. This practice is not only unfair, it is uncalled for. Not every town, city, borough or district has the ability or need to hire a career department. With staffing, benefits, uniforms, station conversion to accommodate a career department, salaries, medical benefits and coverage for days off, converting from a volunteer agency to a career department can be very expensive. Recently, while speaking with a young career firefighter, I was surprised to hear that he was similarly ostracized by his former colleagues. It seems that his fire department needed to hire additional members. At the time, this man was serving as a volunteer. He followed the hiring protocols and was eventually hired. Once he began his career, the volunteers stopped talking to him. While there was no name calling involved, the volunteers distanced themselves from their colleague. The new career firefighter believed that the volunteers stopped speaking to him out of envy or jealousy, since many of them were desirous of the position as well. Regardless of the cause, volunteer and career firefighters need to do a better job of getting along.

Additionally, in large-scale incidents, the paid crew on the apparatus is a set number, and that number is not going to increase during the initial critical stages of the incident. Recall of off-duty firefighters, should that be the preferred method to bring additional crews to the incident, is very time consuming. Volunteer departments are sometimes able to get several more responders on-scene shortly after the alarm is struck and initial response crews leave the station. In times of crisis this is an area where we need better cohesiveness.

For example, if a career firefighter is working inside a structure fire and the conditions degrade with a partial collapse trapping the firefighter, a call for assistance from others on-scene will be made. The firefighter declares a "mayday," alerting all on-scene to their predicament, and other designated response firefighters rush to his or her aid. Once the firefighter in distress is located, it will not be asked what agency he or she is with. Those firefighters who found him or her will not refuse assistance if he or she is from a career department, and they are volunteers. We can reverse the roles, and have a volunteer firefighter in need of rescue, and a career crew doing the locating. The notion that volunteer or career firefighters would even ask that question is simply ridiculous. I have seen career personnel poke fun at volunteers, call them scabs, and laugh at them;

refer to them as *wannabes* and the like. When asked if they ever become trapped and a "scab" came to their aid, would they ask to see a union card, or decide to wait until a union member finds them? Obviously, the answer is no. Then why must we wait for a tragedy to unfold before we can just have mutual respect for each other?

Another argument that is very common between career and volunteer firefighters revolves around training. Certain career members believe that volunteers are untrained, and that the training received (if taken) by career personnel is superior. Blanket statements that contain the words always or never are not very accurate generally. In evaluating the training taken by responders, you must look at individual response agencies, the training taken, the qualifications of the instructors as well as individual firefighters themselves. Many career firefighters are very well trained and are excellent representatives of their profession. The same can be said of many volunteers. Conversely, there are some career firefighters who have only the basic required training, and make statements such as, "I don't need to train; I do this for a living." These types of statements reflect poorly not only on the member making them, but on the profession as a whole. The same can be said of volunteer members, who go only to mandated training sessions, and then make statements such as "I'm a volunteer; what are you going to do, fire me?" Again, this thought process reflects very poorly on all volunteers, even if they do not share the same philosophy regarding training.

In fact, we cannot definitively say that either volunteers or career firefighters are better trained. What can certainly be recognized is that we all can use more training. Nobody can claim to know everything or to not need to participate in training classes and activities. Recently, at a fire service class given locally, a previous student of mine came up to me during a break and said that it was weird seeing me in the class as a student and not teaching it. My reply was that everyone should continually participate in training classes, and that the goal should be to realize that you should never stop the learning process. Career or volunteer, the message regarding training needs to be the same.

Who's in Charge?

Another area where the fire service has serious issues between career and volunteers is the issue of incident command. Often career firefighters indicate they will not listen to, or follow orders given them by volunteer fire officers. The prevailing thought process here is that a paycheck makes someone superior to someone who performs the same job, but does not get paid. Please again allow me to stress that these are isolated incidents, and not every member of the fire service can be lumped into these scenarios. Several years ago, a neighboring fire department was converted from volunteer to a new, mostly career department. The district hired eight full-time, and twenty-four part-time firefighters. The part-timers, me included, covered the overnight and weekend tours. We worked eight-hour shifts, with a set, rotating weekly schedule. One day, while I was working covering a day shift with three of the newly hired career firefighters, I was assigned by my captain to serve as the acting officer. My shift-mates that day were all career firefighters. I overheard them complaining that the "part-time guy" was going to be in charge for the shift.

During the early part of our shift, I gathered all of my crew together for a little discussion. Since I was unfamiliar with the crew as a whole, I asked each of them to review the worst structure fire they had been to. All three stated that they had never been assigned to a working structure fire, either as a driver or member required to make entry. It was at this point that they were confronted regarding the career versus part-time issue. They were advised that the reason the captain put me in charge was that I had prior experience both working at and commanding working structure fires. They all then reluctantly agreed that it was a good decision by the captain to leave me in charge for the shift. Again, the thought process evident here was that one of them should be in charge, solely because they were career firefighters, while I was not. Their own experience compared with mine was of no matter to them, only their status versus mine.

Each state makes its own regulations as to the requirements needed to become an incident commander, and each will enact those required standards. Provided that individuals meet the criteria, they should be entitled to the position and the authority and responsibility that come with it. Volunteers often argue that they are on duty all the time, and career firefighters are only on duty during work hours. With regard to being an incident commander, the number of total hours put in by an individual is of little consequence. As stated earlier regarding firefighters, there are both good and bad officers from the career and volunteer ranks. As was argued regarding individual firefighters, fire officers need to be evaluated as individuals, not lumped in groups by agency or type of service provided. To refuse orders from a competent, well-trained, experienced fire officer simply because that person is a volunteer or a career member is absurd. Firefighters need to stop the childish behavior, and work together, regardless of where each is from, and how he or she got to the incident. One thing that needs to be noted is that both career and volunteer firefighters can be labeled *professional*. Professionalism refers to the manner in which people conduct themselves, the manner in which they conduct their duties and their commitment to their duties. You will note that there is no distinction between career and volunteer members regarding professionalism. A paycheck alone does not make someone a professional. Both career and volunteer firefighters can be labeled as professionals. Professional conduct and attitude, along with competence and devotion to duty is what makes this determination, and we must not insult those who volunteer for any organization and refuse to recognize that they too can be called professionals.

The inability of career and volunteer firefighters to work peacefully together has been an issue for many years, and will probably remain for many more to come. Generally, this uneasiness between the ranks does not interfere with the daily operation of any particular firefighting agency. However, in rare instances, this inability to work together can become a serious and potentially dangerous situation. For example, in 2009 in Florida, tensions between career and volunteer firefighters resulted in a fire station's closing its doors for forty-eight hours to allow tensions to settle down. Mutual aid fire companies were requested to answer all calls for service during the two-day period when the fire department was shut down. The city manager of the town of Minneola decided to close the station, citing safety issues resulting from tensions between the volunteer and career staff members.[1] This inability of crew members to function as a team can be very dangerous. Additionally, this type of problem causes the public to lose faith in the service that they can expect from this agency. While some members of the public will be understanding and even sympathetic to the issues between the career and volunteer ranks, others will not. This type of behavior and the resulting actions taken by the city manager will cause severe danger to the reputations of those firefighters involved for a long time.

Law Enforcement Issues

The same issues about not getting along can be said of law enforcement, but the career members versus the auxiliary officer are not generally as bad as in the fire service. Where agencies have auxiliary or "special" officers, it is usually with good reason. Auxiliary and special officers generally handle mundane, less hazardous duties, which frees up certified officers to handle calls for service and the more dangerous or complex calls. Special and auxiliary members will provide traffic control at religious services, school-crossing services, motor vehicle statute enforcement if so authorized, parades and other civic functions. As with their fire brothers and sisters, many of these people are doing this job as a way to get their foot in the door for any future job opportunities. These people are not trying to take a full-time position away from another officer, but rather, are there to assist in a support function.

Possibly, a reason for tensions or hostility may be that their presence takes away a chance for other members of the department to work overtime. What must be realized is that sometimes the leaders of many agencies must cut overtime in order to stave off layoffs or firings, or to remain within the financial restraints imposed on them by the hiring authority. Therefore, auxiliary and special officers provide a mechanism to help administrators accomplish that goal. Police are often referred to by certain segments of society as 5.0s. The police sometimes refer to their auxiliary and special members as 2.5s or "half-cops." Again, this is unfair, as these officers are trying to do what they can to ease the burden on other officers and the citizens. They generally pose no threat to veteran officers and even handle some of the jobs the veterans dislike. Yet there is still animosity toward them. As with the firefighters, if an officer is in need of assistance and an auxiliary member arrives, would the officer requesting assistance turn them away? Again, the answer is probably not, but until then, these auxiliary or special officers are targets for abuse and ridicule from the other officers.

Issues Involving Strategies and Tactics

Sometimes emergency service providers cannot get along because they simply do not share similar operational or tactical philosophies. Some agencies have an aggressive nature, while others are more laid back. Police have targeted "interdiction" for certain offenses or activities, which they heartily embrace. A neighboring town may not see this as a priority, and will not invest the same resources to solve that problem. But does this mean that one agency is right, or superior to the other? Of course not, but sometimes that is what it seems to be to those looking from the outside in.

As with fire responses, some agencies are much more aggressive regarding interior fire attack than others. Others are more passive and will opt for alternate strategies if life safety is not a contributing factor. However, we will always have those who insist on throwing stones and challenging decisions that were made in the heat of battle. These decisions are often criticized long after the incident, and with the benefit of hindsight. Statements such as "you guys are weak," or "you're no good" are phrases that are often heard by Monday-morning quarterbacks. If the chosen and implemented tactics are different from those another agency would employ in a similar situation, there is a possibility that the decisions will be subject to severe criticism from those in different agencies. This logic makes no sense, but statements similar to these have been made. Regarding the ability to get along, one issue is separate from the other. If an agency doesn't like your strategy or philosophy, they certainly will not want you giving their personnel orders at an incident. Here again, the inability to get along can seriously hinder operations, and may hinder successful resolution at incidents. So long as the strategies and tactics are safe and pose no additional risk to the responder, they should be followed. Only unsafe or unlawful orders should be refused.

Another area where we need to improve relations is again among police officers from other agencies. These references do not apply across the entire police spectrum, but rather only to certain officers. One of the most dangerous tasks for law enforcement is motor vehicle stops. Each year, many officers are ambushed or killed at these "routine" contacts. On many occasions, I have stopped to ask other officers if they need assistance at these stops. The officer asked will sometimes accept the offer and sometimes not. When they refuse assistance, I often wonder why, and then am reminded of my days spent at the police academy. All recruits were taught that there is a deadly phenomenon in law enforcement called "tombstone courage." The thought process regarding this tombstone courage is that an officer can handle the situation alone, and to ask for or accept assistance is a sign of weakness. Sometimes, though, the reason officers refuse an offer of backup is that the person offering the assistance is from another agency, possibly one with whom there is an existing adversarial relationship. Does it really matter who is coming if you are in need of backup? If you are in a fight on the street and another agency officer passes by, would you expect him or her to say "that's not one of ours," and continue on? Will the officer ask to verify that he or she works with you before you accept assistance? Of course not, that scenario is also ridiculous. Yet this type of foolish behavior remains, and we still cannot just get along.

What about Our 911 Dispatchers?

There is another group of people with whom we cannot seem to get along. These are our dispatchers (Figure 3.2). Their job is to field calls from the public, dispatch the appropriate type and numbers of resources, relay information, monitor our well-being and direct additional requested resources to our scene. Yet, we sometimes give them little or no respect. After all, they are "only" dispatchers. That is a title for them, but they are always there watching out for us. These are the people who will reach out for the cop on the car stop who hasn't called in for several

Figure 3.2 Public safety telecommunicators or 911 dispatchers have a very difficult job. They serve as a vital link by providing additional information for responders en route to service calls. Often, they are expected to be intimately familiar with all aspects of their response areas even though they remain indoors and may be charged with ordering additional assets to the scene to mitigate emergencies. They are relied upon to monitor our well-being at calls and often are charged with multitasking. These professionals are entitled to much more respect than they are afforded by those they strive to protect. (Photo courtesy of Middlesex County, New Jersey Sheriff Joseph Spicuzzo.)

minutes to be certain that the officer is safe. These are the professionals who are making calls and dispatching additional resources to your location to assist you. They do a very difficult job under stressful and very trying circumstances, yet they do not always get the respect to which they are entitled.

I saw a television show that recounted the events of a horrific bank robbery and shootout in Los Angeles. During the broadcast, they were interviewing the 911 dispatchers who were working that day. What a marvelous job they did, fielding multiple calls for officers down, units responding and relaying critical information to officers in the field. What was very telling, however, was the emotions that remembering the day brought back for them. They recounted hearing all of the injured officers requesting medical attention, listening to the fear in their voices, yet were unable to actually provide a service to them. Field responders often take for granted what the dispatchers are going through, being intimately familiar with what is going on and being unable to respond themselves to assist. Just doing what they do for us is enough responsibility, but this seems to be forgotten. This is another perfect example of co-workers doing a great job to assist those at the incident. Responders from all fields need to realize what a great service these dispatchers and operators provide to us, and treat them accordingly.

Most emergency responders will tell you that they are not in the business for fame or glory. In fact, you are more likely to face injury, death or liability than fame. Responders seek employment opportunities or volunteer their time for a variety of reasons. Some do it for the financial compensation, pension and benefits. Others choose to join for the chance to help others, or for the "rush" and still others do it because family and friends do it. So, most responders have basically noble reasons to get involved in their chosen discipline. They all share a common desire to have a positive outcome at each incident that they respond to. Given the pure reasons to get involved, why then do we still have issues getting along?

There Are Times When We Do Get Along

There are many times when all responders are able to get along. Whether interdisciplinary or interagency, we often bond in the face of adversity. If a police officer is killed in the line of duty, firefighters will attend the funeral, and vice versa. Firefighters will console each other if one of their own is injured or killed. If we see a particular incident hit close to home, we are there for each other. Other responders will sense this too, and offer a pat on the back, or other form of encouragement. Emergency services also have a tradition of helping each other out, whether it is financially, with resources, or donating blood or whatever is needed. Fire departments that are retiring equipment while it is still serviceable will often find a needy department and donate the equipment to it. I have seen this first hand on many occasions. When the chips are down, we can manage to put aside all our petty differences and work for the common good. We have all been in the shoes of those who are hurting, and we do what we can to offer assistance. After critical incidents, we often turn to other first responders to talk with. We want to be around others who have had similar experiences and we find a way to do so. We will spend many hours listening to someone who is going through a tough time if we have already been there. Responders can be among the most caring and compassionate people in the world. Yet, they can be the most uncaring and callous people as well.

While it can be conceded that we cannot demand that certain people like other people, there needs to be an ability to find a common ground. Responders will rely on each other during times of crisis. The deputy may not like a certain member of a first aid squad, but he will certainly be glad when that responder arrives to relieve him at a CPR call or a serious motor vehicle accident. It is amazing that one can choose to have selective amnesia regarding personality disputes when assistance is needed. If we can put our differences aside for a few moments, why can't we find common ground on a regular basis? I can have mine, and you can have yours, whatever it is that we are fighting over. Why must we always be in competition with each other? Why do we find it so hard to acknowledge the efforts and abilities of our peers and colleagues? If your partner is good at his or her job, why must I try to convince you that my partner is better? Why are we only glad to see each other when we are in dire need of help?

Society today presents us with competition at every turn. We have our choice of cars, beer, fast food, doctors and virtually everything else we want. We explore the positive attributes of the products and services we like the most and purchase the goods or patronize the company based on these evaluations. Companies see fit to compare their products directly against their competition. It seems that those in the emergency service community have done the same thing. My fire engine is newer than yours or we have better guns, etc. While competition can be healthy, it can also be harmful. It can be beneficial if we all aspire to be as good as our leaders, or the "best guy in the department." It can be bad when we denigrate and insult those in "rival" agencies. Once word gets back to a member of an agency that he or she has been negatively spoken about by another agency, there is a tendency for tensions to build between the two agencies.

Using Training to Foster Better Relationships

So why can't responders from all disciplines simply just sit down together and work out the petty differences between them? Some of these issues and differences are so much a part of the various response communities that to try to do so is not possible. Perhaps if they were mandated to train together, they could dispel some of these ridiculous stereotypes and misunderstandings. If agencies are able to train with members of another agency, many good things can come out of the experience. Once we see the need for each other, and we accept that we can help each other prior to a large incident, we can begin to make inroads. I do believe that there is a basic respect for the jobs we each do, and if we trained together, we could expand upon the basic, mutual respect. We sometimes feel that we know more, or can do it better than other responders. Once we accept that there are those who have trained and honed their crafts, as we did, things will begin to get better. Once we get to know each other on a personal level, rather than by associating each other with an agency, the ice will be broken, and things will improve among responders.

There will always be some tensions among members from the same—and different—agencies. Being competitive by nature, responders will generally be of the opinion that their techniques, skills and protocols are the best. There is a belief that the supervisors from their agency are better qualified and trained than their neighbor's. This may be true, but it may not as well. By training together and seeing others as people, and not as members of another agency or department, we can forge better working relationships. Sometimes listening to a member of another department with an open mind can be extremely beneficial. If an incident occurs and responders are sent to assist a neighboring town, the responding personnel will probably be taking direction from the leaders from the requesting agency. If there is tension or dislike between the agencies, there will be distrust and unwillingness to follow orders. By training as a team, the ability to make independent judgments of each other will present itself. Notions that our neighbors are not as good can be dispelled, and when we are called upon to work together, trust will be present. This is what we need to be doing.

Recognition that there are other responders in the world with good training, education and experience is a good start. If we do not train together, we will be unaware of the strengths and weaknesses of our response partners. The fire service relies on assistance from neighboring departments very often, and trust and confidence is often found between them. Other response agencies need to have this same type of working relationships. Towns need to expose all of their responders to their neighbors in a training environment. This will allow for better cooperation between agencies when they are called to work together at an incident. We need to stop the bickering, and we also need to start training and working as a team, and not just as a group thrust together for a short while.

When a major incident occurs, as noted earlier, law enforcement is usually first to arrive. They will call for assistance from the appropriate response agency and await their arrival. If we have a large, unexpected incident, such as a terrorist attack, we will all be forced to come together. This major incident will have the effect of requiring great assistance, cooperation and trust among all of the responders. We cannot wait for these large, complex incidents for everyone to be on the same page. This is not the time to be checking union cards. We need to put aside petty differences, and begin working together for the common good. Jealousy over budget amounts and tools and equipment owned and used by others in our response discipline has no place, if we are truly trying to be problem solvers. Work with what you have, and work harder to get better equipment. As the saying goes, "don't be hating." However, sometimes we do, and when we do, it reflects poorly on all of us in the service. The time has come for us to accept the fact that we need to be on the same page. If we are truly public servants, and we truly desire the best for our citizens and each other, then we must grow up. We must seize the opportunity to get to know one another as people, and not just as firefighter, medic or cop with a certain agency. We have been lucky to have been given time to prepare to work together and train together. I pray that when the time comes for us to put this training to use, we do not fail. We as a response community need to be better than that. We cannot wait for another wave of patriotism to overcome us before we look past our differences and do what we are trained for. In the end, we usually get that wave of patriotism only after a tragedy of epic proportions. Even sadder is that the good feelings are short-lived, and we quickly revert back to our old ways. There is no time like the present to begin to resolve our differences. Once we accomplish this, then we can truly all get along.

Endnotes

1. Roxanne Brown, "Minneola fire department reopens." April 18, 2009. www.daily-commercial.com/041709minnfire.

Chapter 4

Fire and Hazmat

Hazardous materials technicians and firefighters share much of the same training and constantly work on their intradisciplinary training, but they still need to be better trained. The firefighters in our communities are what we might call "jacks of all trades." Generally, firefighters have the broadest amount of training, and will venture into the fields of hazardous materials, emergency medical services, local gas or electric companies, building construction and law enforcement. All these areas cover the things firefighters are called upon to react to once dispatched to a scene. Police and fire crews often request hazmat technicians at incidents, and, upon their arrival, they bring a specific skill set to the incident. Fire crews are trained to take *defensive* actions at hazardous materials calls. These actions include a host of options, none of which involves actively trying to stop the incident, be it a spill or leak. For example, fire as well as law enforcement are trained to: *R*ecognize the presence of hazardous materials; *A*void contact with them; *I*solate the area and secure it from civilians; and *N*otify other, better trained responders to respond to the scene. This acronym is RAIN, and it is used for reporting and responding to hazardous materials incidents. We will explore the fire and hazmat responders in this chapter, the areas in which they can be cross-trained and the issues faced by each group.

Firefighters' Basic Training

The fire service community, like law enforcement, is a very well-trained group of responders in their own discipline. I have referred to firefighters as "jacks of all trades" in the response community for several reasons. First, firefighters get their basic training designed to enable them to respond to and suppress fires and mitigate certain other emergencies. Then, they receive training on wildland fires, vehicle fires, electrical emergencies, carbon monoxide incidents, basic hazmat operations and CBRNE operations. In addition, they get CPR training along with incident command system NIMS training. After their basic training, they are able to respond to and address various emergencies. While in the field, they will assist police with traffic control and diversion, and will often assist the medical responders with first aid, patient packaging and CPR. Where civilians are trapped in confined spaces or in high-angle situations such as window-washer emergencies, fire crews will be called to rescue the victim. They will usually be at the scene of a hazmat release, occurring either in the form of vapors or liquids, prior to the arrival of the hazmat responders. Firefighters are cross-trained initially, but only to very basic levels. Finally, they have the largest vehicles of all of the response communities. In these vehicles, they carry equipment to handle various types of calls for service, and in fact, will choose certain vehicles for certain responses. These vehicles can provide water, aerial devices to reach high into the sky, go into wooded areas, provide lights, extricate trapped passengers from damaged automobiles, etc.

Fire Engine Companies

Fire engines (Figure 4.1) are vehicles that will respond to incidents involving burning of any material in one form or another. Under NIMS, we will eventually "type" all of our responder equipment and apparatus. Typing involves the listing of certain minimum performance standards and equipment that must be on a certain piece of apparatus. For example, fire engines have pumps, hoses, water tanks and assorted other equipment such as SCBA, etc. An engine that is considered a Type 1 Engine may, for example, require that the pump be rated to flow at least 1,000 gallons per minute, carry a minimum of 1,000 feet of five-inch hose, have at least three preconnected hand lines and have an onboard or booster tank that holds a minimum of 500 gallons of water.

Smaller engines, such as "attack engines," may have smaller-rated pumps, carry less hose and less water. Still smaller engines, such as brush trucks, will have even smaller-rated pumps, carry no five-inch supply hose and less than 300 gallons of water. These other resources can be type three or four engines.

The duties of the firefighters assigned to work as crew members on an engine include pump operations, suppression and water supply. The driver of the engine, often called a *chauffeur*, almost always performs pump operations. The training to become a pump operator is fairly rigorous, involving a great deal of mathematics and related calculations to allow the generation of the correct engine pressure to firefighters waiting for water. Additionally, most agencies require that chauffeurs complete a course of instruction on safely driving the apparatus. The pump operator is the crew member responsible for ensuring that water from hydrants or other water sources is directed through the pump and hoses to crews fighting the fire. Engine companies also usually lay supply hose from fire hydrants, which are then connected to the engine to provide a constant water supply for the fire operations.

Figure 4.1 A modern fire engine used to suppress or extinguish fires. It is equipped with a water tank for use prior to having a water supply from a fire hydrant. It has a pump to send the water through the hoses and carries hoses of varying lengths to accomplish different tactical assignments. Many first responders do not make the distinction between fire engines and fire trucks.

Finally, engine company crews are charged with locating the seat, or base of the fire, and extinguishing it. The techniques of suppression differ, depending upon the type of fire encountered. Additionally, the engine is the apparatus that will normally be sent to other specialty-type calls. The engine will be sent to assist the hazmat technicians at spills and releases of regulated materials. Also, those agencies that will still assist with the removal of flood water, and water from broken water lines in homes will dispatch engine companies.

Fire Rescue Companies

If there is a motor vehicle accident, a person trapped in a confined space or a person in need of assistance well above ground level, a fire engine would not be the apparatus of choice. A "rescue" or in some cases, a "squad" would be the vehicle best suited to respond to those types of situations. These vehicles are equipped with the necessary tools and specialty equipment to safely handle special calls such as these. Not all fire departments have rescue trucks, and not all fire departments can handle these specialized types of calls. Rescue trucks can also provide search and rescue, or provide assistance as a rapid intervention crew (RIC) or a rapid intervention team (RIT). A RIC is a team generally consisting of a minimum of four trained and equipped firefighters who stand by at fire incidents where other fire crews are in immediately dangerous to life and health (IDLH) environments. These include structure fires and other calls where exposure to the vapors, smoke, poisons, etc., will cause injury or death if not for PPE. The purpose of the RIC is to be ready for immediate deployment inside an IDLH environment to locate, assist or remove firefighters who become lost, disoriented, trapped or injured.

The RIC operation itself is a very dangerous task. Once the RIC is deployed to assist a crew or crewmember, another RIC must replace the team that was activated. In fact, once a RIC is activated, a prudent IC will request two more RICs to stand by. The concept of the RIC is also relatively new in the fire service. Rescue company personnel can provide assistance at all of these types of calls and perform the various tasks associated with each. These crew members assigned to rescue companies are given advanced training in very specific and technical areas in addition to the basic academy training. Many of these new skill sets require recertification or continuous recorded and documented training to maintain the various certifications.

Fire Truck Companies

Fire truck companies are different from engines and rescues. Truck companies come in several forms, such as ladders or "straight sticks," "buckets," or "platforms" which are elevated to reach heights that ground ladders cannot reach (Figure 4.2). Basically, they are aerial devices that do not carry pumps, hoses or water. Truck companies come in various ladder or platform heights. Some are as little as fifty-foot devices and others can extend to over one hundred feet. Most truck companies do have a *waterway* through which water can be delivered by an engine. These elevated platforms or ladders are called master streams, and when the ladder or platform is extended upward, they are referred to as *elevated master streams*. In general, the duties of truck companies include forcible entry, ventilation both vertical and horizontal, and search and rescue. Ventilation is the process of opening existing features of a building to allow for the introduction of fresh air, or to allow for the removal of heat and smoke. Vertical ventilation may involve opening of scuttles and skylights, or it may involve cutting a hole in a roof. When this task is undertaken, it is to release heat and smoke, to provide better visibility for search and rescue, and to make operations in the IDLH environment a little safer. This is the main reason we see firefighters on roofs cutting holes during structure fire operations. Horizontal ventilation usually involves opening or breaking of windows to release heat, smoke, and poisons. As indicated in the introduction to this book, ventilation, when not done properly, can increase fire volume and spread, and can trap other working firefighters. Fires need oxygen to sustain them, and if we introduce oxygen via ventilation at the wrong place or time, the effects can be very dangerous to all working fire crews. There are assorted techniques used to accomplish ventilation, but those specific techniques need not be reviewed here.

Figure 4.2   This fire truck is a one hundred-foot *"straight stick."* The firefighter will climb this ladder to conduct a variety of tasks, including ventilation and victim rescue. This apparatus carries a multitude of ground ladders, but has no pump or hose and carries no water. If the ladder is extended up, the truck will have one hundred feet of reach. A fire engine can pump water into the rear of the truck into a *"waterway,"* and the water can be used to suppress fires. The waterway is a large pipe that runs under the ladder and has a nozzle on the end. When used in this manner, this apparatus is called an elevated master stream.

Forcible entry is another area where truck companies are generally assigned to work. This involves the firefighter's using a variety of tools to force open a locked door or other means of entry so that other firefighters can enter. Forcible entry also includes padlocks, trunks of vehicles and security gates covering doors and windows. Finally, basic truck company work involves search-and-rescue functions. Rescue companies can serve in this capacity as well. Search and rescues are dangerous operations for firefighters. Generally, the *primary search* begins upon arrival and starts at the fire floor, working up and away from the seat, or location, of the fire. Generally, this task is undertaken without the protection of a charged hose line. Primary searches are quick, cursory searches looking for victims of the fire. Once the fire is under control, *secondary search* operations are undertaken. These searches are much more complete, thorough searches. Often, victims are located during the secondary searches. Further, truck company operations can also include motor vehicle extrication, RIC and elevator rescue.

Fire "Quints"

The last major piece of fire apparatus we will discuss is called a *quint* (Figure 4.3 and Figure 4.4). The quint is a relatively new type of apparatus in the fire service. It is, however, an excellent tool that performs many different functions. A quint is a combination of a ladder company and an engine company. Basically, the quint will have ground ladders and an aerial device, carry some water and hoses and contain a fire pump. This apparatus can be used to pump water to crews fighting a fire inside a structure, while simultaneously extending an aerial device for ventilation or rescue. As with trucks and engines, the size of the aerial device and capacity of the booster tank and performance of the pump are variable, depending upon the specifications requested by the purchasing agency. If properly staffed, a quint can

Figure 4.3   This apparatus is called a *"Quint."* It combines the attributes of fire engines and trucks in one apparatus. It has a water tank and fire pump, carries hose, has ground ladders and an aerial device. Firefighters can place this unit in front of a building on fire and do engine and truck work, including victim rescue, all from this one piece of equipment. (Photo courtesy of the Middlesex County, New Jersey Fire Marshall's Office.)

Figure 4.4 This apparatus has many names, including: "tower ladder," "bucket," "elevated platform," "tower" or simply "truck." This particular tower has a vertical reach of 102 feet and is also a Quint, as denoted by the fire pump and hose. It carries a full complement of ground ladders and 250 gallons of water as well. Note the "bucket" over and in front of the cab of the truck. (Photo courtesy of East Brunswick New Jersey Fire District #1.)

perform engine and truck company operations at the same time, thus helping to resolve the fire much more rapidly.

Brush Trucks and Wildland Fires

Brush fires are smaller versions of the wildland fires we see each year in California, Florida, and the Pacific Northwest. Many agencies have *brush trucks* (Figure 4.5), which are designed to go into wooded areas and other off-road areas. These trucks, when staffed by fire departments, generally have crews that range from two to four firefighters. If properly equipped, brush fire companies will wear a much lighter version of PPE than firefighters fighting structure fires. Not all fire departments own brush trucks, and as you can imagine, they are very rare in urban cities. Traditional fire engines, or pumpers, are generally required to be stationary, with the fire pump *engaged* in order to pump water. Most, if not all, brush trucks have smaller pumps mounted in the rear of the truck. These smaller pumps work independently of the transmissions of the brush trucks. Therefore, brush trucks can pump water while the truck is in motion. In this technique, water can be deployed by the crew member while the vehicle is covering larger affected areas. Other specially trained

Figure 4.5 This older-style brush truck is used to combat brush and natural vegetation fires. The unit has four-wheel drive and carries a winch to assist if the unit becomes stuck. It carries 200 gallons of water with a chemical fire-fighting agent mixed into the tank The truck can be driven and water pumped simultaneously. A crew member rides behind the driver and uses a hose to deploy water. The cage around the body protects the truck during operations in heavily wooded areas.

firefighters support these types of firefighting operations, as well as larger wildland type fires. The U.S. Forest Fire Service and state-level agencies with the same responsibility support local and county resources at these types of fires. These agencies are regulated via federal or state statute, and either provide support or will oversee the actual fire suppression of these types of fires. These firefighters are very well-versed in the use of the incident command or incident management system, and in the case of wildland fires, can manage thousands of responders working on these incidents.

Specialty Units

Excluding specialty types of firefighting such as airport firefighting and shipboard firefighting, I have provided a brief synopsis of the current fire service in the United States. As stated, volunteers make up the majority of the firefighters nationally, but most large cities have exclusively career departments, and other jurisdictions employ their fire protection providers as well. As was noted in Chapter 2, UASI regions throughout the United States have been alotted additional grant monies for training and target hardening. The fire service has also received portions of this funding. It, too, has purchased tools and equipment including fire boats, deluge fire suppression systems capable of flowing 10,000 gallons of water per minute, and response vehicles. Also, some areas have targeted training to assist the USAR task forces. Additional training is offered to firefighters and their agencies in support of the USAR mission.

There are limited numbers of USAR components across the nation, and since they are spread out, response times can be very prolonged. The thought process in training additional firefighters is to assist the USAR task force, which arrives on-scene first. This is very specialized and technical training involving specialized equipment. It would not be practical to train all firefighters in these areas for several reasons. First, most of the USAR training needs constant refreshing and practical training. Not all firefighters, especially volunteers, may be able to meet all of the training requirements. Second, many fire agencies do not own the specialized equipment required, making training and refreshing the training even more difficult. Therefore, it is wiser to train only those with a genuine interest and desire as well as access to the equipment.

Training Differences and Standards

Basic training of the fire service, as with the law enforcement community, varies from state to state. The topics included, number of hours spent on each topic, as well as total number of hours that will constitute basic training, are subject to the discretion of each state. The fire service is fortunate to have a National Fire Academy located in Emmetsburg, Maryland, and this facility offers free training to firefighters from all across the United States. Further, the NFPA is an agency that supports the fire service by producing and updating fire service standards. While these recommended practices are not laws, they provide a framework for national certification in all areas of firefighting.

The NFPA standard for Firefighter I is referenced in NFPA 1001. Most states recognize this standard when training their new firefighters, but, as indicated, the total number of hours for basic training varies by state. Eventually, NIMS hopes to standardize the firefighter training series so that, across America, all firefighters will receive the same training curricula covering the same total number of training hours. Until then, we will be forced to rely on the systems currently in place, which are not standardized. This system, as with police training, leads to each state's having its own standards, and certified firefighters relocating from state to state are generally required to complete the basic firefighter program in its entirety. This lack of standardization leads directly to a lack of reciprocal training agreements, which ends up hurting agencies in need of volunteers. I plan to move to another state upon my retirement, and I have a strong desire to remain a firefighter. However, I do not think, after close to thirty years' experience, that I will have the desire to repeat a course of basic fire instruction. Younger members of the fire service who move to other states may share this attitude as well. Basic fire training is a long process that takes weeks of full-time or months of part-time attendance. With young firefighters needing to work second jobs, or having children or a working spouse, the time needed to be retrained drives otherwise viable volunteers away from serving. This is a totally unacceptable scenario, and one that needs to be addressed immediately.

Retention of Volunteers and Response Problems

In New Jersey, I was once a part of a delegation that sat down with our congressman to discuss recruitment and retention of volunteers. He realized that we need to find a better way to get people involved and to retain their services. The proposals put forth were unsatisfactory, to say the least. Proposals included waiving local permit fees, or the motor vehicle registration fees of the vehicle driven by the volunteer. This type of thought process is insulting to the volunteers, who donate their time, efforts and talents to their community. They risk their lives, spend countless hours away from family and friends, and the dangling carrot for each of them is a few dollars a year. Those already in the fire and EMS services probably do not need much incentive to stay involved. However, we do need to recruit new, younger members, and we need to be able to keep them as members. Those present at the meeting with our congressman recommended property relief or federal income tax credits. Once these substantial suggestions were proposed, the topic was quickly changed. It seems government wants and needs volunteers, but is not willing to recognize the sacrifices they make to remain active, and to reward their efforts in kind. Mandated training, initial training, refresher training, not to mention the surprise call for service at any day and any time all take their toll on volunteers. Years ago, volunteers were allowed to leave work to respond to emergency calls. Today, most employers will not even consider this request, and most volunteers neither work where they live nor volunteer their services.

Given these factors, what happens when disaster strikes during traditional workdays? Who will be around to answer the call when paid responders are not available? Calls for volunteer service in areas under volunteer service providers' control during "normal" work hours are quickly becoming very problematic. In the county where I reside, I have seen several towns convert to career EMS because there are not enough volunteers to answer the service calls. The biggest issue in making the move to a paid or career service is the daytime calls. In the fire service, there has been a tremendous increase in what we call *automatic aid agreements*. This is an agreement that basically provides that one fire agency will be automatically dispatched to another fire department's area for all fire calls during a specified time. Most of the agreements center around weekday calls for service during the normal work hours of most workers. While these agreements are certainly a good idea, they mask the underlying issue of a dearth of responders, especially during the workday. Sooner or later, we will not have staffing for some volunteer services, and we need to address this issue now. Areas under the direct control of volunteers need to recognize the declining numbers of volunteers, and steps need to be in place when the time comes to remedy the problem. Automatic aid agreements work, but they are akin to treating a broken bone with an aspirin. The pain goes away, but the bone is still broken. Another critical problem is presented for debate and consideration, and hopefully there will be answers and strategies in place when the crisis needs to be addressed.

Issues Faced by Smaller Agencies

Another factor to consider in the cross-training world of fire department personnel is how smaller departments make do with less manpower and resources. In a perfect world, an engine company would have a minimum of four crew members. One would drive and operate the pump at arrival, the officer would direct the operation and possibly assume command of the incident, and the remaining two firefighters would stretch the line and suppress the fire. Similarly, a truck company would have the same crew size and the duties would be similar to their job description. The chauffeur would remain with the truck to control the aerial, the officer would direct tactical operations, and the remaining two crew members would be assigned to forcible entry, ventilation, or search and rescue.

We already are aware that most of America's fire departments are mostly composed of volunteers. We have also already discussed the shortage of volunteers and the issues faced in retaining those we already have. So what happens when the call goes out and you do not get the four-man crew you need? What if you only get three firefighters? Well, you leave the station and work with what you have. Similarly, if you cannot get a truck company out of quarters in a timely fashion, does that mean that the engine company waits at the door until the truck company gets there to force the door? Of course this is not the case. Engine companies carry tools often used by truck companies to be used where there is no truck company on-scene. The crew is trained to know that they will need to gain entry themselves, and they do what is required. So what we see here is that the fire department members are already doing more with less.

Some communities are unable to afford the very expensive ladder trucks and rely on *mutual aid* from surrounding departments to provide the apparatus and crew for them. Mutual aid is different from automatic aid in that it is not automatically sent upon receipt of a call for service. After a need is observed, a specific resource is requested to report to the location at the discretion and direction of the incident commander. Mutual aid is sometimes called for in the form of additional *alarms.* Alarms, in the fire service, are the calls for service and which apparatus will respond to that particular call. Alarms have certain apparatus or resource requirements. For example, a first-alarm fire may require two engines, a truck company and a rescue company. If the incident commander sees a need for additional resources and calls for a second alarm, the same number of additional resources will be dispatched to the scene. Each jurisdiction will set the parameters for their alarm structures. If an incident becomes too large for the resources on the scene to control, additional alarms are struck to bring additional resources to the scene. In large cities, or where there are large departments, additional resources from that department respond. In smaller departments, resources from outside the department are called, and this assistance is called mutual aid.

Smaller departments are often forced to have their crews conduct multiple tasks until such time as secondary apparatus from their department arrives on location, or until mutual aid responders arrive. The use of mutual aid in the fire service is not a new trend. When crews are forced to conduct multiple tactical assignments, the incident becomes much more dangerous for them. If a small department gets a report of a *working* fire and responds with an engine, it is forced to use those resources to start the initial operations. If there is a large body of fire, the crew will be forced to try to control the fire before conducting search-and-rescue operations. If they do not do so, the fire may grow in complexity, and the crew may become trapped with no means to escape. So, the crew gets water from the hydrant, forces entry if needed, stretches a hose line into the house, and knocks down the bulk of the fire. There are no relief crews on the scene for them yet and they are aware of a potential resident's still being inside the home. These crew members are now forced to begin search operations. The SCBA worn by firefighters do not contain a limitless supply of air. So now the crew has used some of the precious air and are forced to begin a new operation, but without a full supply of air. Unfortunately, this is a reality in today's fire service, especially in small towns or where there are small departments. It is a dangerous situation to begin a fresh operation with less than full SCBA cylinders.

Monitoring the Health of Fire Service Personnel

Firefighting is very taxing work, and crews are required to report to *rehabilitation* during the course of their stay at the incident. Rehabilitation units are established to monitor the health of firefighters after they have completed some of their tactical operations. This is especially true after firefighters have worked in an IDLH environment, or after they have used a SCBA cylinder. Some departments will limit the number of SCBA cylinders their firefighters may use during an incident. Wearing SCBA taxes the human body, and safety and medical officers want to ensure continued good health of crews. In my experience, most agencies with a SCBA policy will limit the firefighter to two cylinders. After that, they must be assigned to operations not requiring the use of SCBA, or be relieved from duty. Heart attacks are the leading cause of firefighter fatalities, and rehabilitation allows each firefighter to be examined by a medical professional, usually an EMT. At rehab, firefighters will be stripped of their heavy turnout gear, be given fluids and will have their vital signs taken. After a short time in rehab, after relaxing, the vital signs will be taken again. If the firefighters' vital signs are returning, or have returned to normal, the medical officers may recommend allowing them to return to duty or clear them for another assignment. If the vital signs are still abnormal, the firefighters are held in rehab for a longer period, and the process is repeated. If, after a period of time determined by the medical officers, the firefighters have not made substantial recovery, they are taken to a hospital for further evaluation and definitive medical treatment.

Calling for Specially Trained Resources

There are some responses where technical rescues are needed and even the best-trained fire departments are not qualified to provide assistance. For example, in New York City and at the Pentagon after 9/11, there was substantial catastrophic structural collapse. In cases like these, even on smaller scales, even more specialized responders are special-called to the scene. These USAR teams are very highly skilled teams that are required to meet stringent training and equipment standards. They are well staffed and equipped and trained to respond to a number of technical responses. Once they are able to meet all of the federally mandated requirements, they are evaluated and eventually certified as USAR.

Some components of the USAR include medical, technical rescue, building collapse, hazardous materials and communications. These teams have been sent all over the world after earthquakes and similar national disasters. In 2003 in New Jersey, a parking deck was being constructed for an Atlantic City casino. The structure collapsed, killing and injuring a number of workers. Incident command personnel requested the New Jersey USAR, who responded, took control of the emergency, and worked to resolve the crisis.[1] The Oklahoma City bombing in 1995 also necessitated the activation of multiple USAR elements from different states. These teams, once certified federally, are subject to being dispatched all over America and the rest of the world. There are not many of these teams in every area, and one drawback to them is their response time. They are not full-time members, and once activated, they come from various destinations to a central reporting location. Once there, their equipment is prepared and the team responds to the emergency. In New Jersey, which has a UASI region, support teams are being trained and equipped to assist the USAR. These support teams will provide a faster on-scene time for specially trained responders in advance of the arrival of the USAR.

Hazardous Materials Technicians

Hazardous materials technicians and specialists are among the rarest of the emergency responders. While there are many responders who have been trained and certified, only a small percentage are actually working or affiliated with a working hazmat response team. Many of the hazmat technicians and specialists are often already trained as firefighters. Technicians are given additional training in identifying hazardous materials and substances, as well as CBRNE materials. They have equipment that enables them to make identification of these materials, and other equipment that they can use to mitigate the emergency. Technicians are trained to take offensive measures at a spill or release, which means that these are the responders who will patch leaking drums, trucks or cylinders. They will work to actively control the release of liquids and vapors, and will assist in cleaning the spill or release once it is contained.

Hazardous materials *specialists* have additional training above the level of technicians, and they will specialize in either road-driven trucks or rail cars. These specialists are given in-depth training regarding either or both of these hazardous materials transport vehicles, including construction of the vehicle and techniques used to mitigate spills and releases. In the event of a CBRNE incident, the hazmat teams will be requested on-scene rather quickly. It will be these responders who will set up operations to identify the agent used in the attack. The IC and staff will use this information during the planning and tactical decision-making processes of running the incident.

Generally, these teams work very methodically and deliberately, taking great pains to assure the safety of their team members and the public. These responders are sometimes required to wear Level A fully encapsulated suits to conduct their operations. These suits are very physically demanding to wear, and hazmat teams begin monitoring the vital signs of their members prior to donning the suits. Further, the hazmat team will establish a decontamination line prior to allowing its members to enter the hot zone. The decontamination line is set up in the warm zone, and is in place should an emergency occur.

At an incident, there are three zones: hot, warm and cold. The hot zone is where the issue in need of mitigation is located, and responders working there are required to wear specific PPE. The warm zone, often referred to as the contamination reduction corridor, is where decontamination occurs. The cold zone is an area outside the other zones, one that is completely safe for all those present without any PPE. The decontamination line is very important, because any victim must be decontaminated prior to being taken to the medical area for treatment and transport for definitive medical care. Regardless of the medical condition of responders or victims, they *must* undergo decontamination prior to triage and treatment. As indicated, hazmat teams, like bomb squads, are a rare resource and will be stretched very thin in the face of a major incident. Politicians must ensure that these resources are properly funded and staffed to be able to respond to terrorist activities. As noted earlier, without their expertise, we would be unable to determine the agent being used and when the atmosphere is safe for us to breathe.

Hopefully, you now have a basic understanding of fire department and hazardous materials team's apparatus, crew needs and duties performed by various responders. This review is just that, a basic overview of the fire service, including the hazmat technicians and specialists. With the basics laid out, we can now explore the specific areas where we need to better train our firefighters and hazmat team members. We will also explore how this cross-training will benefit both the responders and the public they serve.

Additional or Continuing Training

An area in which fire department personnel can be better trained is terrorism awareness. The fire department personnel in this country respond to different parts of their response areas, and for different reasons. Fire department personnel, for example, routinely handle suspicious-odor calls. Here is an example of why terrorism awareness is a good training area. Urea nitrate was used in the World Trade Center bombing, which took place in 1993. A substitute ingredient for the urea is urine. Therefore, should a fire department be called to a scene and notice a strong odor of urine, this *may* be an indication of a possible bomb-making facility.

Fire department responders may be called to respond to small fires or even small explosive incidents coming from inside buildings. Again, this *may* be an indication of terrorist activity. There are other possible indicators, such as lack of furniture or a tremendous number of unredeemed coupons. These coupons indicate the potential for fraud. Unredeemed coupons can be sent to manufacturers as though a customer had redeemed them. The fraudulently obtained funds are then used to support terrorist groups abroad. Fire department personnel who do not have access to terrorism awareness training will miss the boat regarding some of these possible indicators.

Should signs of terrorist activity be present, law enforcement should be immediately notified and fire department personnel should refrain from any type of intervention. Similarly, hazmat teams may be called for certain spills, odors, or releases, and they too should be trained to report any of the terrorist indicators to law enforcement, again without any attempts at intervention. Generally, these activities occur in larger cities and towns, but they can occur in smaller towns as well. Some of the 9/11 hijackers lived in an apartment complex in Passaic County, New Jersey prior to the attacks. While the area where they lived was no small hamlet, neither was it Newark or Jersey City. Providing this terrorism awareness training for fire personnel just puts more "eyes" on the street. The New Jersey Division of Fire Safety (NJDFS) has already begun offering a basic terrorism awareness class for fire inspectors. This is a great place to start, and the response community as a whole should investigate this training for all responders.

The fire service is, in many ways, the most versatile and best trained of our emergency responders. They carry the biggest assortment of tools and equipment. They are able to enter IDLH environments during fires to conduct life- and property-saving missions. They carry tools and equipment for specialty rescues, motor vehicle extrication and first aid emergencies. They carry cones and road flares to assist law enforcement in traffic control, and have some basic equipment to set up defensive operations in advance of a hazmat team's arrival. Hazmat teams support fire agencies by bridging the gaps where firefighters are not trained to enter, or where they may have to control spills and leaks. Hazardous materials crews generally do not carry hoses and water to support their mission. Firefighters supply these pieces of equipment to the hazmat teams.

Generally speaking, the fire and hazmat teams are up to the task on a daily basis. They are able to handle the routine calls for service and can even handle the unusual calls. But what will happen if we are subject to another terrorist attack? New York City has among the largest police and fire departments in the country. Even with all of their paid, on-duty resources, they were immediately overwhelmed on 9/11. This is certainly no criticism; they did an excellent job under conditions that are difficult to even imagine. But, what if this event on a smaller scale took place elsewhere? We have already deduced that police responders will be the first to respond to the scene. Fire and EMS are sure to be close behind. Depending on the type of attack or incident scenario, other resources may be called to respond as well. We have acknowledged that those in the fire service are very well trained. But, they are not skilled in all areas, and there are certainly areas where they will need to be exposed to better training.

Terrorism Attack

As an example, we will choose a chemical attack. For this hypothetical event, the target is a very large, enclosed shopping mall on the last Saturday before Christmas. The time of day is just after noon, and the mall common areas, stores, and parking areas are all full to capacity or near capacity. Suddenly, in many areas throughout the mall, people begin choking, coughing, suffering from tearing eyes and are passing out. The mall is in a state of panic and chaos as the shoppers fight to exit. Many victims collapse at various exits shortly after making it to fresh air. Security personnel at the mall call 911, and the local police, fire and EMS are dispatched.

Many in the response community believe that the fire responders are equipped to deal with this. However, there is a very large fly in the ointment. The structural turnout fire gear worn by firefighters is not rated to protect them from chemical warfare agents. These agents will eventually defeat the garments worn and the firefighters will become victims as well. Also, the SCBA used by many fire departments are not rated for CBRNE exposures. Again, the breathing apparatus may be compromised and the firefighters will become victims. In discussing this fact with firefighters who have had some training, I have found that most are unaware that their gear and SCBA are not designed for use in chemical environments. We need to get them into training classes to ensure that they can protect themselves. Our fire officers need to be aware of these limitations as well, since these officers will be directing incident strategies and operations. Once they arrive on an incident, they feel compelled to act, as do the police. Failure to realize the restrictions on protective gear can be fatal. Unless chemical protective ensembles with CBRNE-rated SCBA are used, we are potentially exposing our responders to the same agents as the victims, even though they may believe they are protected.

Continuing with this scenario, have we trained our firefighters and hazmat teams on what they can expect next? Once the firefighters are on location, they will call for the hazmat team to respond to the location. As noted earlier, these team members will be critical to the safe and effective response to, and resolution of, the incident. As the incident unfolds, injuries become more and more apparent, responders continue to respond in to help, and the hazmat team will begin setting up decontamination facilities. Firefighters are trained to understand the reasons and need for the various types of decontamination, but they may *not* be trained to assist in the actual decontamination. They will be charged with establishing the water supply for the hazmat team, and may utilize their truck or engine companies for *gross decontamination.* This type of decontamination occurs when victims are simply flushed with water to remove the majority of contaminants from their bodies or clothing. But different decontamination techniques are used for different exposures to certain agents. Steps may include blotting, wetting and removing, or may include wetting and removal, etc. Firefighters are generally only minimally trained to conduct gross decontamination.

If victims are overcome from exposure to the CWA, they may be unable to put themselves through the decontamination process. These victims are subject to *emergency decontamination.* This entails victims' being put on a backboard and placed on a roller or other type of transport system. The backboard and victim are pushed through a line of water streams, with the goal being to remove as many contaminants as possible. Victims unable to walk for various reasons also need this type of decontamination. If there are a large number of victims and only the hazmat team members are trained in decontamination, there will be significant issues trying to get all of the victims through decontamination. At this point in the operation, ambulances and paramedic units are coming to the scene, and other resources are being *staged* at an off-site location for use later in the incident. The nearest hazmat team has arrived and begun preparing crew members to take air samples and to try to determine the type and name of the agent and concentration of same that was used in this incident. Further, the team has selected a large area and begun setting up the decontamination line so that victims can be decontaminated and released for treatment.

Shortly after hazmat begins accepting victims at decontamination, the victims clearing the decontamination process are being treated and eventually transported to the hospitals. Hospitals are quickly going to a condition known as *divert,* which indicates that they are unable to accept any more victims of this event. Hospital staff, doctors and nurses are being requested to respond to a makeshift medical treatment area near the incident, where they can assess and treat the victims of the attack. Chaos is running rampant; accountability of victims and responders alike is getting troublesome. More and more responders are arriving, with little or no guidance from those in charge of incident mitigation. The scope of the incident is beyond the capabilities of local responders, and mutual aid from all surrounding towns and even from neighboring counties is responding to the scene.

Police on the location are trying to restore order and help establish what they call perimeters. Access to the site is becoming difficult, both for responders entering and ambulances leaving with victims. The IC has slowly gotten a grip on his or her response, and accountability and staging areas are getting established in accordance with incident command protocols. All responders are doing what they can within the scope of their training. Command is advised that the hazmat team is getting high-level readings, indicating that the agent at the mall is sarin. This is the same agent used in the Tokyo attacks in 1995, where thousands were injured, and approximately one dozen were killed.

Problems Associated with Secondary Attacks

Suddenly, and without warning, there is a massive explosion coming from one of the areas where the victims are being taken for decontamination. The hazmat team members there who were assisting with victim preparations for decontamination, doctors, nurses and firefighters are injured; significant numbers of these have life-threatening injuries. There are also dozens of dead bodies resulting from the blast. We will expand upon this incident later, after we address several issues in this incident to this point.

Where was the security to prevent the explosive attack? Who was watching for potential indicators of a secondary attack? Sadly, during the chaos and confusion of the response, no one was on *heightened alert.* We need to expose our firefighters and hazmat teams to more training regarding terrorist attacks. We need to explain methodologies and current trends. Law enforcement gathers intelligence and *classifies* it. They share this information among other law enforcement agencies in the area that may be affected by the reported intelligence. But has anyone notified others in the area that may respond to an emergency there? Most likely they have not.

Fire and hazmat crews need to be trained in the area of suicide bombers and current trends involving IED, which is a disturbing trend currently seen in the Middle East. Law enforcement seminars are quick to provide training in this area, but fire, EMS and hazmat do not always attend the training. We cannot keep life saving training and intelligence from our other first responding units. We need to not only open the classes to these responders, but we need to actively invite, and even request their presence. As noted earlier, no one knows when and where the next terrorist event will take place. Accordingly, we need every responder to be kept in the training loop. It does no good to classify information and then expect uninformed responders to handle an incident. We have exceeded the realm of seeking out and expecting secondary attacks.

Our enemies, knowing that we have given some training in the areas of secondary attack to some of our responders, have upped the ante. They now plan attacks in waves of three, meaning we must now be aware of the possibility of tertiary attacks. We need to keep fire and hazmat crews abreast of the current trends in terrorism. There is another recent trend that I will introduce into our ongoing scenario later in this chapter.

Terrorism response is a fluid, rapidly evolving discipline for which we do not provide all of our responders enough training. We need to, and in fact, can do better in this area. The area of suicide bombers is not the only aspect that will be explored at this time. The other glaring hole in the training aspect of this scenario to this point is the failure of the firefighter to be situationally aware. While we were exploring the cross-training of law enforcement, we demonstrated this phenomenon in regard to the blocking of fire hydrants and parking directly in front of the structure that is on fire. Firefighters are focusing all of their attention on the area of the incident that is their specialty. Since they have received little, if any, training in the area of operations at terrorist incidents, they cannot be expected to be aware of certain factors.

It can be noted how the failure to provide cross-training can impact an incident. Training is only the first step in making all responders situationally aware at all times and at all incidents. The human machine is a very intricate and sophisticated structure. The mind is the single most important aspect of our being. In times of true crisis, we instinctively revert into a *fight-or-flight* mentality. That is to say, we will either engage in the fight of our lives, or we will choose to flee and live to fight another day. People who study the way humans deal with training will advise you that to have an action to be second nature, or done without thinking, you must do that action thousands of times. Police are trained in drawing and firing their weapons over and over, so that when they are in fight-or-flight mode, they draw the weapon, come directly on target, and fire until the threat is neutralized. This training is refreshed each time an officer goes for firearms recertification. Imagine if the police were trained in firearms while on the firing range at the police academy and never again. When there is a genuine need to draw and fire their sidearm, how effective will they be? Would we, as citizens, feel comfortable knowing that the officer may have not drawn his sidearm in many years? Yet, we give the firefighters very basic training, and we never refresh it. How is this fair to these responders? They are not experienced in doing anything other than traditional firefighting duties.

Firefighters will train in all of their traditional duties—suppression, ventilation, extrication, etc. We need to train, or retrain fire and hazmat personnel on trends and indicators of possible secondary events. There have been incidents where firefighters have been targeted, but they have been isolated and indeed, some were unavoidable. During the Los Angeles riots after the Rodney King verdict in the 1990s, there were incidents where firefighters were shot at by snipers. There have been other incidents where suspects have barricaded themselves in their homes and then set the home on fire, and have shot at and killed responding fire crew members. There is sometimes no warning that an attack is imminent, but there are some terrorist warning signs and potential indicators. As with law enforcement responders, fire and hazmat personnel are trained to be very aware of their surroundings at all times. Subtle changes in smoke patterns or color of flames at a fire can mean any number of things. Firefighters engaged in search operations are trained to listen very intently for faint cries of victims, and to maintain contact with walls inside structures being searched to avoid being confused or disoriented. They have had experience and repeated training sessions with these skill sets. Since we revert back to our training under stressful situations, we need to give the fire and hazmat responders a broader base to draw upon.

In our scenario, I did not comment on where the explosion came from, or what the detonation mechanism was. Did you come to your own conclusions? Do you believe that this secondary event was a result of a terrorist bomber? For the sake of our scenario, we will say now that the explosion *was* the result of a suicide bomber. I will tell you that his intended targets were, in fact, the responders on-scene who were rendering aid to the injured. The suicide bomber chose a very opportune time to detonate his bomb, and no one was even aware he was there. He simply presented himself as a viable rescuer, present simply to aid in the rescue efforts. Once the scenario was in a state of operation where the bomber felt the most damage would be done, the bomb was detonated.

Requesting More Resources to the Scene of the Attack

At this point, the IC is stunned, and the scene grows even more chaotic than it was before the blast. More resources are requested to the scene. These resources include additional fire, EMS, and hazmat crews. The resources, which were in staging, are moved up to the mall and the responders begin sorting through the carnage, seeking survivors of the blast while continuing the treatment and decontamination of initial victims. The newly arriving responders get a short briefing on what has just happened. These responders are tense, nervous and frightened. They wonder if there are more bombs present. The fire and first aid responders realize that they have no real training regarding terrorist attacks. They push their fears aside and continue to work, feeling somewhat secure. They do recall during that the training they received, what seems like ages ago, referred to secondary attacks. Feeling relieved that the feared secondary attack has occurred; they are able to focus a little more on their tasks at hand.

The terrorists have had everything they planned go exactly as they have envisioned. There are multiple fatalities, and hundreds of casualties at the mall. But they are still not satisfied, and they planned for a third wave of their attack. Responders are trying so hard to concentrate on their duties that they are completely unaware that recent trends in terrorism indicate likely tertiary attacks. This is only one area where they are unaware of the recent trends, and things will go from horrible to whatever comes after horrible.

We move along in the incident to one hour past the suicide bomber's strike on the mall. We now have regrouped, we are getting the decontamination process moving much more quickly and have triage working feverishly. A second hazmat team has arrived, and additional medical professionals have responded. Again, as before, there is a sudden explosion not too far from the main entrance of the mall, where there are still hundreds of victims awaiting decontamination and medical evaluation. This explosion is the result of a large car bomb, and it wreaks more havoc on the already tumultuous scene. We will terminate the scenario at this time, and move forward.

While this scenario may seem to be a little on the extreme side, it is not. Current trends in terrorism show a desire for secondary or tertiary events at a single incident. Further, the more disturbing trend is multiple simultaneous attacks, a scenario that is nothing new. In fact, we have seen it since 2004, at the latest. In 2004, Al Qaeda planned and executed coordinated simultaneous attacks at three hotels in Jordan. In Amman, the Grand Hyatt hotel, Days Inn, and Radisson SAS hotels were all victims of an explosive attack. During this attack, 57 people were killed and approximately 110 wounded.[3] These attacks have generally been credited to Abu Musab Al Zarqawi, an Al Qaeda leader who was later killed in Iraq by American military forces.

This same type of simultaneous attack was seen in London, England in July, 2005. There were three suicide bombings on the subway at approximately 8:50 a.m. These suicide attacks were followed approximately one hour later by a suicide bombing on a city transit bus. Fifty-two were killed, and as many as 300 were wounded during these coordinated attacks.[4] Accordingly, terrorists worldwide have seen the concept of multiple simultaneous attacks at work. From a terrorist viewpoint, they were very successful.

Bombings require a full response of all emergency response assets. When one area is overloaded, the request for emergency responders cannot be met and the consequences are magnified. We need to ensure that our firefighters, who will be working at the first of these attacks, are equipped and ready for other attacks. Fire crews, along with all other emergency responders, need far better information regarding current and evolving trends in terrorism. Terrorism training should be all encompassing, including small and large international groups and domestic groups. In both of the real-world scenarios, which I cited to support my argument, suicide bombers were the delivery system of choice by the terrorists. These weapons systems are truly the world's best *"smart bombs."* These bombs are detonated when and where the bomber decides. Suicide bomb attacks are prevalent in many areas of the globe. While we here in America have not had to deal with the issue of an actual suicide bombing, can we be certain that one, or more, is not coming? Success by the overseas terrorist groups may result in copycat incidents at home. Most police officers and other law enforcement personnel may not even be trained fully in the area of suicide bombers or responding to an incident. If law enforcement is lacking, almost certainly the fire service is lacking in this area as well. Unless there is a leak or source regarding the identity of a bomber and the intended target, there is virtually no way to stop one of these attacks. Therefore, we need to train all first responders in this area, and the time for this training is now.

What If We See Attacks Similar to Those Seen Overseas?

If we are attacked in a fashion similar to London's and Amman's, will all of our responders be ready to react? When the incident is finally mitigated and we begin the recovery and healing process, Americans will demand to know who is to blame. Our law enforcement personnel will provide this accounting. Any attack site will be a crime scene. We need to ensure that our fire, hazmat and EMS partners do all they can to try to preserve evidence. I was fortunate to take a class on incident response to terrorist bombings. This class was provided to my classmates and me in New Mexico, and the federal government paid for all of our expenses. During one part of the class, we were detailed to look for *"evidence"* of the bombing. The tiny pieces of the "bomb" and its components, which we surrendered to our instructors, were amazing. In the aftermath of a bombing, we need to let our hazmat and firefighters know how tiny the evidence of the event may be, and what steps they need to take to preserve this evidence. Notice that I am referring to actions taken *after* the victims have been treated and transported away from the incident site.

In a prior chapter, we explained why training on IEDs was so vitally important. Firefighters and hazmat crews are also in need of this type of training. IEDs come in many shapes and sizes, but fire and hazmat crews need to have some training, at least to the awareness level. Security forces in malls, airports, hospitals and other sensitive areas are trained to be vigilant for possible sources of explosives. Fire and hazmat crews will be in the hot zone rendering aid to victims and conducting other tactical operations after an attack. These responders need to be aware of possible indicators of the presence of an IED. Again, since the term IED can be broadly applied, I do not suggest that any responder can safely announce that there are no IEDs present after an incident. However, we can train them on where to look and what to look for. Hazmat technicians are among the most deliberate and safest members of the response community. They will not rush into any incident without gathering as much information as they can. We can all learn from the skills demonstrated by these workers. If called upon to act, they will inspect the area prior to taking offensive actions. We need to take this diligence and add to it by providing training on suicide bomber tactics, as well as IED placement and tactics. The more eyes we have looking out for each other, the better off we will be. But, the eyes need to know what to look for. Currently, there is not enough training for us to feel confident that these responders have the best available and most current information.

Additional Areas for Cross-Training

Currently, many firefighters and fire service recruits are given CBRNE training at the academy level. Hazardous materials technicians are even better trained in the CBRNE field because they will be charged with taking offensive actions after an incident. We have addressed the CWA aspect with regard to the firefighters. As with police and other law enforcement responders, biological attacks will provide few immediate tactical responsibilities for fire responders.

One area where we need to better train our fire crews is in the realm of radiological attack. Most firefighters get a small block of instruction in radiological incidents. We have already explained what a dirty bomb is, and other aspects of radiological attack. Most fire agencies in America are completely unaware of the types of radiation present every day, how to detect it, and what the readings of radiation mean. Further, the same lack of detection equipment is prevalent in the fire service. I would propose that fire agencies be provided with training on radiological incidents above the small block of instruction they currently receive. A radiological and nuclear awareness course is available at the local level from the federal government, again free of charge. We will address the training later in this book.

Hazardous materials teams already possess the equipment to detect radiation, as well as the skills to interpret the readings on their meters. But, as already noted, there are far too few hazmat teams in place in America. Fire crews need access to these detection tools as well. In my county, we have periodic meetings of all of the fire chiefs in the county. At the first such meeting in 2009, the county fire marshall announced some of his plans for our portion of the Homeland Security grant funds. One of the initiatives proposed was to try to equip each piece of fire apparatus with one radiological detection pager. This would be a great place to start. The fire marshall also announced that the fire academy was hosting several of the aforementioned radiological and nuclear awareness classes. This will be an excellent program, if his office is able to purchase and provide the equipment and training. It is this type of a forward, proactive approach to training and equipping our firefighters that needs to be duplicated all over America. Radiation is similar to carbon monoxide, in that it is odorless, colorless and tasteless. It can be detected, but not without detection equipment.

Another area in which we should have all of our firefighters and hazmat crews trained in is first aid, SAED, and CPR. While career fire departments and hazmat teams may require that these skills be maintained, some volunteer agencies have no requirement, rather leaving this skill set to the first aid and EMS communities. Fortunately, where both EMS and fire protection are provided by volunteers, a great many professionals maintain certification in all of these disciplines. As indicated, the fire service is sort of a "jack of all trades." It will be responding to any large-scale incident involving terrorism, but also it responds to other, less involved incidents on a daily basis. The fire apparatus is the response vehicle, which generally provides the most manpower. On almost every fire apparatus, there is some form of medical equipment. If we can marry the first aid resources with trained EMTs on the engine or truck, we can provide a shot in the arm for the overworked EMS resources.

We have pointed out that there are currently not enough hazmat and EMS providers. If we could train the fire crews in first aid, it would be of tremendous assistance. While I understand that not all firefighters would be willing, we need to train as many of them as we can. In some instances, where BLS crews are unable to respond, the fire department is called upon to assist. This apparatus is then referred to as a *first responder apparatus*, usually an engine or rescue. The main drawback to this type of response is that the apparatus cannot transport the victim, and there may be a reluctance to treat a victim by firefighters. Since we have an identified problem with first aid responders, perhaps this is an area where we can fortify a weakness with another response asset. It will be conceded that the added training burden to maintain both EMS and fire certification may be too much for many responders. However, if we do not explore the possibility, we will not know about the interest in it. Training our hazmat team members in EMS is also something worth looking into, but as there are already too many tasks and not enough teams, the fruits of this endeavor may not be bountiful.

Summary

We have explored several areas where our firefighters and hazmat technicians can be better trained. These emergency responders spend a great deal of time honing their particular skill sets to be ready to respond to a variety of calls for service. In general, these two response disciplines provide some of the most in-depth and well-rounded training programs found in the emergency response community. However, as with the police being under trained in the areas of fire response and hazardous materials response, fire and hazmat are equally under-trained in certain law enforcement skills. While it is understood that there are certain aspects of law enforcement training best left to the law enforcement community, there are still a great many areas where our fire and hazmat responders could be afforded training opportunities. Some of the areas where we can do better in training fire and hazmat teams were discussed in this chapter. While it is conceded that we cannot have all responders certified in all response disciplines, we can provide a better base of training skills. The need for terrorism awareness training was suggested, and a strong case can be made for the value of this training for fire department and hazmat personnel. The training will not require release of classified information, but rather will serve as a platform for raising the awareness of terrorist indicators for newly trained responders. The more responders in our communities looking for potential indicators of terrorism, the safer we will all be. This awareness training can be conducted in a short period of time, and no special equipment will be required to use it on the street. Career departments can integrate the training into their existing training program. Volunteers as well can use existing training time to participate in the requested training classes.

We have seen how we need better training in recognizing the indicators of IEDs and how they will be used against us. Radiation is an area where not only our responders, but also most of our citizens, are completely under-educated. We need to provide awareness training to all responders and make more in-depth training available to those who desire it. Our emergency responders, both career and volunteer, spend many hours in training each year, in maintaining certification, or simply maintaining proficiency in their chosen field. It is unfair to demand additional training for our responders to operational levels across disciplines, but we need to at least encourage the process.

In this chapter, we provided a scenario of a terror attack at a hypothetical mall in America. While the depth of the attack and the callous manner in which it was produced may seem unlikely; we must be prepared for something along these lines. Remember, prior to 9/11, who among us really ever thought that commercial jet airliners would be used as weapons of mass destruction? Who would have thought that these "weapons" would be potent enough to cause the collapse of one, let alone both of the World Trade Center Towers? Our enemies and terrorist groups have decided to explore new terrorist mechanisms and delivery systems. They are open to previously un-thought-of weapons of mass destruction.

In our hypothetical scenario, I was able to document two other similar attacks that occurred in England and Jordan. Some of the terrorists we face are very well educated—scholarly, even. These death mongers study proven tactics and seek out new and untested techniques to be used against us. If these individuals decide to bring the war to America, we will need every available responder to be ready, willing, prepared, and able to meet the disaster head-on. To accomplish this bold and aggressive challenge, we need to all be on the same page. We must take every opportunity to cross-train with our first responder partners. To resolve the next attack quickly and with as little loss of life as possible will require that all responders have access to and take advantage of every training opportunity that is presented. The cost, in terms of loss of human life, may be too much to overcome, if we are not prepared. We will have no one else to blame but ourselves. Let's take advantage of the Type A people around us, and challenge them to expand their basic training options and be truly ready to act. Other countries around the world have already taken some of the measures that I have proposed, and it is time for us to get around to providing access to the information, training and equipment that the fire and hazmat responders need to protect us. They are there for us, and we need to be there for them. Our very lives may depend on the training they get.

Endnotes

1. WFTD.com "Firm reviews temporary closing of Minneola Fire Dept." July 18, 2009. www.wftv.com/news/19787227/detail.html.
2. CNN.com/US: "Four killed in NJ parking garage collapse." October 30, 2003. www.cnn.com/2003/US/Northeast//10/30/garage.collapse/index.html.
3. Hassan M. Fattah and Michael Slackman. "3 Hotels bombed in Jordan; at least 57 die" New York Times online, November 10-2005. www.nytimes.com/2005/11/10/international/middleeast/10jordan.html.
4. Alan Cowell. After coordinated bombings, London is stunned, bloodied and stoic. *NY Times* July 7, 2005. www.nytimes.com/2005/07/07/international/europe/07cnd-explosion.html?r=1.

Chapter 5

Emergency Medical Responders

On any given day in this country, there are hundreds of thousands of calls for service for our emergency medical responders. The responders most likely to respond to these calls are referred to as, among other things, EMTs. The vehicles they ride in are called ambulances, rigs, buses and BLS units. Volunteer and paid medical professionals routinely handle these medical calls. Some of these are members of first aid squads, or rescue squads. Still, some of these calls for medical service are covered by law enforcement personnel, and fire department professionals handle others.

In a separate category of medical responses are those that require further medical training than can be provided by EMTs. In cases where the injury or illnesses are more serious, the EMTs on-scene will request better trained and equipped responders. These responders, called paramedics, or medics, are usually referred to as ALS. In many areas, these units will not transport the victims, but the team will split up and one medic will ride in the BLS unit with the patient, and the other medic will follow behind. Additionally, the medic units are trained to intubate victims, provide intravenous lines, perform telemetry, administer certain types of life-saving medications, and even pronounce victims dead. Some states have nurses to be assigned to the ALS units, and they have the authority to act within the scope of their training. In some cases, when the resources to bring victims to the hospitals are not available or are insufficient, we will bring the medical facilities and more definitive care options directly to the scene of a major incident. While this is not a common occurrence, medical facility personnel have been trained to accomplish this mission. Given the limited numbers of BLS and other transport-capable vehicles, this option provides a better mechanism for treatment of multiple victims in cases of catastrophe or disaster, or in scenarios where there are significant numbers of victims.

Issues Facing Our EMS Providers

Emergency medical technicians and paramedics are well trained in their response field and do an excellent job under very trying circumstances. As we have already indicated, there are simply not enough trained personnel, and more importantly, there are not enough transport-capable responders. In some areas of the country, BLS services are provided by contract ambulance services, or in other cases, hospitals will provide both BLS and ALS services. During my years as a responder, I have attended several drills that simulated one form of event or incident with multiple victims. In each of these scenarios, those reviewing the drill noted a severe shortage of EMS response assets. Even when the drills have involved *simulating* responses of medical responders, there was always an issue of a shortage of those able to perform triage, treat and transport the victims. While a great deal of emphasis is placed on treatment of victims, there just are not enough responders available. Even cities and towns that are very proactive will not see the benefit of stockpiling mass numbers of ambulances in the event that a need one day arises. No one would propose that governing bodies stockpile apparatus or vehicles in this manner. To even propose such an apparatus stockpile is ridiculous.

Training for EMS

Medical basic training for EMTs is fairly intensive, and as with fire and police training, varies from state to state. The students are exposed to a variety of topics and are afforded hands-on opportunities to hone their skills, which include the systems of the body, treatment of lacerations, fractures, abrasions, patient packaging, C-spine protection, administration of oxygen, CPR and many other areas. The responders are then able to answer medical calls for service, and are able to properly handle the requests. While there may be some limited instruction in areas other than treatment and assessment of injuries, we clearly are not fully educating these responders. They will be targeted by terrorists, as secondary or tertiary targets, the same as law enforcement and firefighters. Just as in war, the terrorist will seize an opportunity to target a medical responder. Victims of the initial attack will be awaiting the arrival of trained medical assistance to treat their wounds and to transport them for definitive medical care. However, if our medical responders are *situationally unaware,* there is a good chance that they will be potential victims of a secondary attack.

Additional Training for EMS

We have explored areas within the fire and hazardous materials and law enforcement arenas where the training is in need of expansion. We will now review the training areas and classes that should be provided to our medical responders. In this chapter, we will explore all areas of medical services, and not limit ourselves to EMTs and medics. We will touch upon private EMS providers, nurses and medical doctors. While not all of these medical service providers will be called upon in the event of a crisis, in some instances most, if not all, will have a direct impact on the successful resolution of the incident. Broad changes are required, and we can only begin this process if a standard curriculum for these responders is set. Again, I do not propose, nor do I endorse a program where we make all first responders interchangeable. People need to remain in their own response area, and they need to maintain a certain comfort level regarding the duties and tasks they will be asked to complete. This does not exclude or replace the need for comprehensive training across the board.

CBRNE Response Training for EMS

The logical place to begin our review is to revert back to hazardous materials incidents, and follow this up with CBRNE response. It is generally agreed upon in the response community that these are the types of events we will be faced with after our next domestic incident. We will begin, therefore, with hazardous materials responses. We have already explored the instances of TIC and TIM, and how deadly these materials and chemicals can be. But what type of hazmat training is provided to the people charged with the initial triage, assessment and treatment of victims' injuries? Sadly, the broad answer is that we do not give them enough. The Department of Transportation produces a number of copies of the NAERG. These are routinely distributed to all members of the emergency response community. This reference text provided by the DOT is a great tool for the initial response to an incident, but it has certain shortcomings. The guidebook has several different sections that allow the responder to identify a hazardous material or substance. The responder can seek the information by an alphabetical index, or if the identification number is known, it can be referenced by this number. Once the product is identified, the book will refer the responder to the orange section, which is called guides. These guides will provide basic information about the material and provide guidance on what measures need to be taken under certain circumstances. One drawback to the guidebook is that many materials are listed under the same general response guides. We need to provide better information, one wherein specific chemicals and compounds have their own listings. The National Fire Administration (NFA) has such a guide, but most responders do not readily use it. It is used in a similar fashion as the NAERG, but is far more comprehensive. Copies are available on CD, and can be easily loaded into a PC or laptop for use in the field. This is the type of manual, or reference text, we need to be providing to our medical responders. If the information is not completely accurate, treatment options may be ignored, dismissed or improperly applied.

Should we be attacked via a CWA, either a nerve agent or a vesicant, our responders need to be made aware of signs and symptoms of exposure. As previously indicated, exposure to pepper-type sprays and nerve agents share initial symptoms. So, if the secondary attack is set in motion after responders are on-scene, and it is not a suicide bomber, but a nerve agent release, awareness of the signs and symptoms would be of paramount importance. Blister agents will present as a burning sensation, but contact with human flesh will result in the development of fluid-filled blisters. If the agent is inhaled, these blisters may form in the lungs. Regardless of whether the exposure is inhaled or absorbed the mere exposure is very dangerous. Are all of our medical responders familiar with these types of attack scenarios and the symptoms associated with these different agents? Many times, the calls received by our dispatch centers are filled with inaccuracies, mistaken interpretations of events and a host of other types of factual errors. Therefore, crews are dispatched to an incident that may be totally different from what they expect to find upon arrival. If this is the case, have we given our responders enough information to properly and quickly make correct deductions after observing the patient? Once responders exit their vehicles, they may be exposed to the same agent as the initial victims. The jumpsuits worn by many EMTs are classified as Level D protection. Most responders in the medical field are not given PPE which will provide them protection against the various CWAs. Even if the chemical ensembles are available and on the ambulance, if the dispatch information is inaccurate and the crew exits the ambulance into a contaminated environment, they too may become victims.

As indicated in an earlier chapter, firefighters arriving at a hazmat call will plan a route accounting for wind direction and topography. They will stop at a distance and use binoculars to evaluate the scene prior to approaching the area. But, can we say the same of our medical responders? Have we become so complacent that we ignore our own safety, or is it that we have left these responders so untrained that they will exit their vehicle only to become a victim? This is not fair to our EMTs and medics. To provide a short block of instruction on the indicators of chemical attack during basic training and again never refresh the material is unwise and unsafe. If a class on responding to terrorist incidents had been provided in 2000, the current state of terrorism would not have been analyzed and explored. Trends and attack methodologies are always changing, and our responders need to be kept aware.

Biological Training for EMS

The next area where we can expand our current training would be in the area of biological attacks, methods and parameters. Biological incidents take time to surface, due to various incubation times, as discussed in Chapter 2. Hospitals are trained to detect rises in certain types of emergency room visits (Figure 5.1). Once a spike in a certain type of illness is observed, actions are undertaken, and protocols are followed by hospital staff to determine whether there is an epidemic, or if there is some other root cause for the dramatic increase in emergency room visits. EMS providers are in

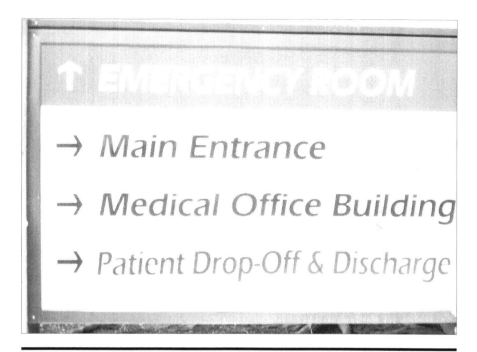

Figure 5.1 Hospitals and their staffs will be vital after a terrorist incident. Responders need to be familiar with decontamination principles and protocols. Hospital workers, including doctors and nursing staff, need to be trained in these and other areas in order to be properly ready and able to treat victims of a terrorist attack.

a position to observe these same trends in the field. If they observe a spike in certain calls for service, they can begin to alert hospital officials, who can then track the information and make an independent determination regarding the illness. More importantly, if they are aware of symptoms of certain predominant biological agents, they can take better and more appropriate self-protection measures. The medical communities are very well aware of *universal precautions* to be taken when treating patients, yet they sometimes do not follow these recommendations in the field.

Universal precautions include eye protection, a mask over the nose and mouth, gloves and clothing to ensure there is no exposed skin. If these precautions are not taken and an airborne agent, such as plague, is present, upon entry to a patient's residence responders are subject to becoming affected by the same agent as their victim. Here again, it may be complacency, or it may be a total lack of under-standing of one's environment. It is possible that responders do not wish to intro-duce an aura of fear into the residence of the patient seeking medical assistance. Additionally, it may just simply be that until a scenario wherein these precautions are needed presents itself, responders are willing to take the chance. Medical responders are not alone with this attitude. Firefighters will take off their SCBA during the overhaul phase of a fire operation, feeling that the toxins and poisons present have been removed since the fire has been extinguished. In reality, carbon monoxide, hydrogen cyanide and other toxins are still present, and firefighters do sustain lung injuries due to complacency and bad practices.

Radiological Training for EMS

The next area in our CBRNE arena for medical responders is radiation, or radiological attacks. The medical response community in America has a severe need for training in this area. Medical responders are intimately aware of the need to decontaminate certain victims of certain incidents prior to allowing them into ambulances (Figure 5.2). Decontamination, when done properly, can substantially reduce the potential of *secondary contamination.* This phenomenon occurs when a victim has been exposed to a substance, and the substance is transferred to the responder. That responder has therefore become contaminated. The responder, during the course of treating the patient, also becomes contaminated as a direct result of contact with the victim. If a victim is decontaminated prior to triage and treatment, the medical responder can focus on treatment options rather than worrying about secondary or *cross contamination.*

As indicated, medical responders are well versed in the practices and need for decontamination to avoid secondary or cross contamination. But, are they aware of

Figure 5.2 Decontamination is a must before patients who have been exposed to certain chemicals or agents can be transported via ambulance to hospitals. Everyone needs to realize that regardless of your injury, you MUST undergo decontamination once exposed. Failure to do so will contaminate not only those treating and transporting the victim, but the ambulance as well. (Image copyright Loren Rodgers 2009, under license from Shutterstock.com.)

how victims of an RDD are decontaminated? As indicated, alpha, beta and neutron radiation present as particles of an unstable atom. Without training and detection equipment, decontamination will not occur, because responders will not know that the particles are present. Further, radiation offers no cues for responders to detect via their own senses. The EMTs and medics need to be familiar with the steps involved in decontamination radiation, as well has having training in detection of same. Hazardous materials technicians generally perform decontamination, but firefighters have recently been undergoing training as well, in an effort to support the hazmat teams. Firefighters' training is generally limited to gross decontamination, which would not be used for decontamination of these particles. The more responders who are aware of the potential for radiation being present, and who can assist in decontamination, the better off we will be.

Radiation decontamination requires a *survey meter.* A trained operator of the meter will slowly pass the robe of the meter over the body of the person being decontaminated. Based upon readings on the meter, radioactive particles are located and then removed with duct tape. Again, this technique is used only for the alpha and beta particles. There is no decontamination for gamma radiation, which are waves. If an ambulance is dispatched to a small explosion with injuries, will they be thinking this may be a dirty bomb? They will probably be thinking more along the lines of a fireworks-related incident, or maybe a pipe bomb that was detonated and caused traumatic types of injury. I am not trying to promote paranoia, but terrorism is not something that will disappear simply because we refuse to talk about it. Medical responders take pride in their equipment, and take great measures to ensure its readiness for the next call for service. If they are unaware that they may be using a contaminated vehicle, they will be the cause of cross contamination, which is simply unacceptable. If we are planning to attempt to outfit fire apparatus with radiation detection equipment, and law enforcement personnel are expected to be so equipped, we must complete the chain and equip our BLS and ALS responders. Medical transport units are already in short supply, and we cannot afford to have any of them taken out of service due to their being contaminated by radiation or any other substance.

Explosive Incident Training for EMS

The last part of CBRNE response that needs to be addressed is the area of explosive incidents. Explosions produce a wide variety of injuries, and most medical responders are able to treat the injured. Explosions produce a shock wave, which will mean many serious internal injuries. This is also referred to as a *shock wave.* These blast injuries include, but are not limited to, the eardrums, abdominal area, lungs, heart and other internal organs. The initial shockwave is capable of causing death without any evident physical trauma. The other, more obvious effects of explosive injuries are traumatic, resulting from shrapnel. While the medical responders are more than adequately trained to handle the medical issues associated with explosive incidents, they would be better served if they had additional training about explosive incidents. Specifically, training should include awareness regarding suicide bombers, techniques used abroad to target the responders and tertiary attack potential.

As already noted, most responders are trained to be aware of possible secondary attacks, where they are the preferred targets of the terrorists. However, we must now be alert for a possible third event. A tactic used by terrorists is to dispatch a suicide bomber to target the responders—*all responders*—after the shock of the initial attack is reduced, and responders are focusing on treating the injured. The tactic that is very successful entails having the suicide bomber dress as a first responder, and medical scrubs are the most easily obtained of the responder uniforms. We have indicated that medical resources will be in great demand but in short supply during mass casualty scenarios. Having a secondary event occur is of great value to the terrorist, for a variety of reasons. Dressing the suicide bomber in scrubs, which are commonly used in the medical profession, will allow them immediate access to the inner areas of the incident, where responders will be feverishly working to treat victims. Once there, the suicide bomber can choose the time to detonate the bomb he or she is carrying that will be most opportune for maximum deaths. This will further reduce medical response capabilities, while increasing the numbers of victims in need of medical attention. Providing training such as incident response to terrorist bombings and suicide bomber awareness would be a good block of instruction for the medical responders. These blocks can be added to basic training, or can be made available for continuing education units (CEU), which are mandated by many states for responders to maintain their current level of certification. Investing a little time and effort may yield great benefits when the training is used at an actual incident.

Private EMS Providers

There is another source of medical service providers who are almost always completely overlooked until a catastrophe or mass casualty incident (MCI) occurs. These medical professionals are the private EMS providers. There are a great many of these privately run and owned transport services throughout America. Generally, the personnel who staff these private ambulances are required to hold the EMT certification, or equivalent, for their state. In the event of any MCI, these private citizens will undoubtedly be called upon to assist local medical service providers. However, since these EMTs are most likely untrained in any form of CBRNE response, terrorism response, attack methodologies and the like, they will not be aware of additional dangers that may be present. These additional responders will play a huge role in the ability of incident managers to transport more victims for definitive medical care in the aftermath of any incident. It would be fitting to include these responders in training classes along with emergency responders from all disciplines. In states of emergency, all unnecessary travel will be suspended, and these ambulances will be requested in unusually high numbers to assist local EMS providers. The owners of the services will be asked send their resources to assist in whatever manner they can. The incident managers will certainly put their medical skills to great use to treat, and eventually transport the victims to area medical facilities. Since these private service employees will be thrust into the middle of the incident, we owe it to them to be as prepared as possible.

These private employees will need a minimum level of training to be adequately aware of the dangers they may potentially face. While the private transport providers are not ordinarily dispatched to emergencies, they do have the requisite training for such events. Some private service providers are used as first responders in some areas, and will have better base training and an understanding of situational awareness. Those EMTs who are providing a transport service will most likely need to be provided supplemental training. For example, these responders may be required to establish a rehabilitation post during an event. This assignment requires many activities to occur at the rehabilitation location, and the transport service providers may not be aware of what is required of them to complete this task. They will need training as to what to expect, as well as what to provide at the rehab post. Additionally, these new responders will need to be trained in a host of other areas.

Training for Private Resources

Since these transport service providers are unaccustomed to emergency response issues, many of them should be provided in-depth training. They should be trained in the principles of incident command, or incident management, including NIMS. These responders will be working within the framework of these management systems, and need to have at least a basic understanding of the principles contained therein. They will be subject to different radio procedures and protocols, may be tasked to a strike team or task force, or be assigned to a staging area. Further, they will need to be versed on unity of command, span of control and other rudimentary aspects of the IMS in place. It would be wise to provide this type of training prior to an incident's occurring, since all responders will be operating within the same system.

As noted, these responders, if generally assigned to transporting patients to and from medical appointments, are redirected to the scene of a MCI, or terrorist event, there may be gaps in their training. Another serious area would be in the procedures and importance of decontamination. We have already addressed the value and importance of pretransport emergency decontamination. If the experience of the EMT is solely to pick up and deliver patients to medical appointments, they may have no idea that these MCI victims may contaminate the ambulance or the responder. If the incident management team requested these private medical service providers, there must be a genuine need. To lose any of these vehicles or personnel as a result of poor, or nonexistent training may have devastating effects on the medical operations in progress.

Lastly, regarding the training of the privately contracted medical services provider, is the overall CBRNE response training. These medical personnel should be receiving the same training that I proposed for all medical service providers. They have the same issues regarding CWA signs, symptoms and attack methodologies. These employees may also be targeted by terrorists via secondary attacks or devices.

The real issue regarding the training of these responders is twofold. The primary issue will be one of payment of these responders while they are being trained, as they are employees of the service provider. Private employers may be reluctant to pay their personnel for training that is outside of their normal job responsibilities and duties. While one could argue that the training may help save additional lives, possibly even the lives of the employees themselves, the employers may be very reluctant to provide payment.

The second issue is who will provide training, and to what levels. The training may be provided at basic training, but this can vary greatly. Private companies will provide the requisite medical skill training, but may not delve into terrorism response, or even emergency response training. They will probably spend a great deal of time on the internal policies and procedures of their individual companies. So if not at basic training, where do we refer these responders for training? We will address this issue in greater detail later in this book, but fire academies, local training academies or institutes and local vendors may be able to provide some of this training. Law enforcement or fire personnel may be able to supplement the balance of the missing training, and may be able to provide it free of cost to the private provider. Again, we are not trying to propose that all responders be inter-changeable, but we should get basic awareness-level training out to as many poten-tial responders as possible. Awareness of odors, signs and symptoms and similar training will provide these additional, nontraditional responders with a better understanding of the issues and challenges they will be facing. Putting these hastily acquired helpers in the field with no training can be detrimental if they are unable to handle the assignments given to them.

These private companies are a great, untapped resource for incident managers during any terrorist attack or MCI event. We will be forced to rely on them if there is a genuine need, but we also need to prepare them for what will be expected of them. A small investment of time and money should yield great rewards if an MCI or terrorist event were to occur and these people are called upon to provide assistance.

Paramedics, Doctors, and Nurses

The bulk of the training we have discussed for the medical responders in this chapter has focused on the EMTs who are in the field and private EMS providers. We will now explore training proposals for paramedics, nurses and doctors. In the event of an MCI, there is always the possibility for medical personnel from hospitals to be brought directly to the scene. While they would be "out of the way" of the responders on-scene, they will be very close to the incident. Many hospital person-nel have been trained on the techniques and practices in making the hospital "field ready." When emergency rooms become filled to capacity and there are still victims at the incident location, the hospital will send out staff to man a *field hospital*. The military or National Guard, who have training in this area, will be of great assistance in setting up these field, or mobile hospitals if they are already on-scene. While the concept of bringing the medical staff to the wounded is not new, and is very sound, we need to be sure that these medical responders are aware of the situations into which they are being thrust. While we do not need to make the responding medical staff from hospitals first responders, we can certainly acquaint them with several aspects of the types of attack scenarios to which the victims may have been exposed.

Should a field hospital be set up after a CWA incident, having the on-site medical responders aware of attack methodologies and dissemination methods could be of great assistance to them in performing their lifesaving duties. Medical doctors, nurses and paramedics should already be acquainted with the signs and symptoms of nerve and blister agents. This training will ensure that the victims are subject to the correct medical countermeasures. Bringing the medications directly to the scene would be of immense value and save many lives due to having victims' getting immediate medication. While the doctors and nurses are able to properly diagnose and treat victims, we need to prepare them for what may occur in the field. Exposure to chemical, biological and radiological warfare agents without training as to what to expect is not fair to these medical professionals. If the doctors, nurses and medics were aware of the possibility that they would be targeted as secondary victims, they would demand training. At the very least, field hospital locations should be swept by K-9 teams with bomb-sniffing dogs for any explosives, and the location should also be equipped with chemical-agent and radiation-detection equipment. While the security aspects of setting up a field hospital will generally fall under the purview of law enforcement or the IC, these responders may have other issues to deal with, and may overlook the need to ensure that the field hospital is established in a safe area. If the personnel who will be assigned to work there are well trained, they can suggest, or demand, that the area be properly inspected prior to transporting victims to the location, and well before the facility is staffed.

Decontamination and Hospital Security

Decontamination is a very important step in the process of victim rescue, triage, treatment and arranging for definitive medical care. The hospitals and medical professionals who work there should be well aware of the need for emergency pretreatment decontamination of victims prior to their arrival at the hospital. Some hospitals are now equipped with decontamination showers at the entrance to emergency rooms. These professionals are aware that if a victims who are contaminated by any agent or substance enter the hospital, those victims will contaminate the hospital, as well as any staff members they have contact with. Logic would dictate that hospitals with decontamination facilities train their staff on the use and importance of the decontamination station. In cases of emergency, doctors from hospitals without decontamination facilities will certainly be requested to assist at these field hospitals. These nurses and doctors should be trained on decontamination procedures and protocols to ensure that proper hygiene and treatment of victims is maintained at field hospitals.

Hospital staff and security members do a good job of securing and monitoring the radioactive isotopes stored in hospitals as part of their nuclear medicine departments. Doctors and nurses should have access to additional training regarding these isotopes. Terrorists, both foreign and domestic, who have an interest in radiological attack scenarios will be very interested in isotopes stored in hospitals. As previously noted, an RDD, or dirty bomb, is a bomb that has radioactive materials present as well as conventional explosives. When the container housing the radioactive materials is compromised by the detonation, the radiological materials are released into the environment. Doctors and nurses should be exposed to training regarding radiological attacks and security protocols for protection of medical radioactive materials. Terrorists see hospitals as a rich environment for the gathering of radiological sources for their RDDs. The more trained people who are aware of the potential for theft of these isotopes, the better. As will be discussed later, free training in this area is available, and should be taken advantage of.

Summary

EMS members in America are very well trained. There are different levels of responders within the medical response community. This field represents the largest diversity in training among all of the emergency responders. The basic levels are that of first responders, such as police and firefighters. The next level is the basic level for BLS providers. These responders are generally referred to as EMTs, and are the largest number of the medical responders. The next level of training is that of paramedic, or ALS providers. These responders have more extensive training than EMT personnel, and are able to provide a much greater and medically comprehensive service to their patients.

The next level of training is that of nursing. These professionals have extensive medical training, and can perform additional duties above the EMT as well. Finally come the medical doctors. These are the most highly trained and skilled members of the medical community. For our purposes, all doctors are treated equally, meaning that we will not differentiate between a general practitioner and a specialist. What all of these responders have in common is the ability to treat sick and injured persons. While each has a different level of training, they could all benefit from additional training outside of their traditional field. Incident management training would be most beneficial, as the incident managers will utilize these professionals in emergent situations. Additionally, CBRNE and suicide bomber training will only make them better prepared to respond to an emergency. Hazardous materials training such as decontamination procedures and other relevant topics will also serve to better enhance their response capabilities.

Private transport providers are the most under-utilized medical responders on a day-to-day basis. We need to get them up-to-date training to prepare them to be a great asset in case of MCI events. Spending a little time and effort to acquaint these responders with training across traditional boundaries will only enhance our response capabilities. Medical responders will be the most needed resource at an MCI event, yet they will be the fewest responders. We need to maximize their training and make sure they are capable of handling these challenging events. Due to a lack of BLS and ALS response vehicles, we will undoubtedly request the assistance of privately owned and operated medical service companies. These assets need to have the same training as those who traditionally respond to calls for service in the field. Prompt delivery of medically trained personnel, equipment, medications and transport capable vehicles will be critically important in the event of an MCI. There is a definite need for these responders. This much is certain, and largely accepted by the incident managing community. The need again is identified, and now we must provide the training.

Chapter 6

Office of Emergency Management Personnel

Introduction

The Office of Emergency Management (OEM) is charged with managing all types of situations and emergencies. Under current incident management protocols, the members of OEM will not be charged with commanding the incident, but will be involved in support functions within the IMS. The OEM is present in many different levels of government, but is charged with the same role at each level. Municipal OEM, the lowest level, is charged with many tasks and functions, all completed at the local level. The next level of OEM is at the county level, and again, it is charged with certain responsibilities and tasks, as is the next level of OEM, which is state level. On the federal level, there are various agencies such as the Federal Emergency Management Agency (FEMA), for example, which will have responsibilities for emergency management functions. We will first explain what each level of OEM is charged with, and then we will delve into which areas of training would be beneficial to each level of OEM, and why.

Local or Municipal OEM

Local or municipal OEM is the first level of emergency management that will become involved in any incidents that occur on a local level. These incidents include, but are not limited to, severe weather incidents, fires, flooding incidents, electrical or power disruption, cold- or heat-related incidents, as well as acts of terrorism or hazardous materials incidents. In cases of any incidents or events that impact on a town, borough, city or any other municipal entity, the local OEM is advised, and it begins assisting the incident management team in mitigating the incident or event. Frequently, the OEM staff will make themselves available to assist the IC with tasks such as ordering resources. Typically, each municipal entity is required to have an emergency operation center (EOC) that can become a command post during extended or involved operations at the local level. The EOC is a permanent fixed facility, and generally has certain features that make it amenable to serving as a continuously operating command post. The EOC should have its own power supply, and a backup power supply is desirous. There should be multiple land-line phones, fax machines, photocopier capabilities and a large work area. Additionally, access to Internet providers, cable television and other electronic media should be available, along with ample parking and restroom facilities. Local OEM staffing levels can be as few as a single coordinator, or may have a coordinator and several deputy coordinators, plus support staff. In larger cities, the OEM will have a large staff and considerable budgets and resources. Generally, these personnel are paid city employees who work a full-time schedule. Smaller towns may have a coordinator who is serving on a voluntary basis and may have a tiny budget and few resources.

The responsibilities of local OEM coordinators are somewhat varied. They will report to incident managers, and will provide whatever assistance they can to aid in managing the incident. These services may include arranging lodging for displaced residents who have no place to go, contacting disaster relief agencies such as the Red Cross or Salvation Army, or arranging for other resources to be provided to the incident management team. Arranging for these services is called *Logistics* under the ICS. Logistics section chiefs are tasked with many responsibilities, and will establish all facilities used during the incident, such as the command post, staging areas and any other requested facilities. Further, under the incident command structure there are six *Units* under the logistics section, including the medical unit, food unit, facilities unit, communications unit, supply unit and ground support unit. The logistics chief will be charged with establishing any or all of the units on an as-needed basis by the incident management team. The local OEM coordinator may be called upon to serve as a logistics section chief on incidents at the local level.

The last, and main function of OEM we will discuss is the creation and maintenance of the local Emergency Operations Plan (EOP). This document is required to be on hand and used at incidents or events at the local level. The EOP has a series of chapters referred to as *annexes*. Each annex will cover a different aspect of the overall EOP. For example, the Fire Annex will outline policies for management of fires. It will define the responsibilities of the fire chief and the other members of the department, and will list all equipment owned by the fire department and any special duties or responsibilities to be carried out by the fire department. After this annex is prepared, it is reviewed and signed by the OEM coordinator, and the fire chief. The EOP has various other annexes that are designed to provide guidance for any anticipated incident that may be under the control of the municipality. Some of these annexes include radiation, police, EMS, continuity of government, housing, shelter, annexes for pets, etc. Even very small municipalities will have EOPs that are quite large. Once the EOP is updated at regularly scheduled intervals, copies are provided to the coordinator, deputy coordinators, mayor and EOC, and a copy is forwarded to the county OEM. Many states require that the county approve the local EOP prior to its being reproduced and distributed. County OEM will keep a copy of the EOP for each municipality in its respective county. These EOPs are referenced when the county OEM is requested to provide assistance to the local jurisdictions.

County OEM

The next progressive level of OEM is at the county level, where the same basic duties and responsibilities conducted at the local level are required and provided. At the county level, the OEM coordinator is a full-time, paid county employee with a support staff, deputy coordinator(s) and a line budget. The county coordinator is available to support local OEM coordinators and reviews and approves local EOPs, inspects the local EOC, holds regular periodic meetings, reports directly to the appropriate county government liaison and to the state level EOC for its region. The county coordinator has a support staff that will be available to local OEM coordinators as well as local incident managers. The county support staff will generally be integrated into the logistics section and serve the incident managers as needed. The county support personnel have access to the county EOP, which is authored by the county coordinator. They are able to refer to the appropriate annex within the county EOP and arrange for resources as requested by the appropriate authorities. For example, if there is a large fire in our county, the county fire coordinator will be notified after a fire reaches second alarm. Once the fire reaches a third alarm, a fire coordinator will be sent to the scene. Upon arrival, he or she will check in with the IC, and will provide support to the IC as needed. Generally, these coordinators will arrange for additional resources that may be needed to mitigate the incident. They will also be tasked with ensuring that there is ample reserve fire protection in the jurisdiction where the fire in located, as well as in the towns that sent mutual aid resources to the initial fire.

The county OEM coordinator is required to periodically update the county EOP, in the same manner that the local coordinator is required to update the local plan. This document is very similar to the local EOPs, in that the document is approved and maintained by the county OEM. The county EOP will have the same annexes as the local plans, but will be much more detailed obviously, since it will be responsible for a larger geographic area and serve more residents. The county plan is approved and reproduced, and copies are sent to all county agencies that have responsibilities under the plan. The county coordinator and the leaders of county government sign the plan prior to reproduction and dissemination. Additionally, the county will also have its own EOC and most counties will also have a backup

EOC located a distance away from the primary EOC. This enables the OEM to function even if the primary EOC is in an area that precludes its use during an incident or event. Generally, the county EOC is larger than those of municipalities, but larger cities will also have very large EOCs. The county EOC will have additional seating for such entities as Red Cross, local power providers and county government leaders, and will also maintain an independent radio communications facility. Another aspect of the EOC will be that it will provide a call-in number for the public, in an effort to control the expected rumors that will be circulating throughout the community during a significant incident. This dedicated phone line is referred to as the *rumor control line.* Periodically, the county will be requested to test certain aspects of the county plan, by either state or county officials. These drills are sometimes conducted as tabletop scenarios, and are sometimes hands-on scenarios. After the drill is completed, the deficiencies that have been identified are discussed, and appropriate adjustments are made to the annex that was subject to the drill. The EOP is then revised and updated, and new policies or procedures are identified and implemented. The new EOP is then signed as required, reproduced and distributed as deemed necessary.

State OEM

The next level in the hierarchy of OEM belongs to the state. Each state has its own OEM, often falling under the control of the state police. The state will have OEM officers assigned to cover all of the counties within the state. These specially trained state troopers (if under state police control) will be available to provide support and resources to the counties to which they are assigned. As with the county level coordinators, if the state OEM is requested at an incident, they will provide support as needed by incident managers. The state will not generally come in and attempt to assume command of an incident, but rather will provide additional resources to the incident managers in an attempt to mitigate the crisis. However, if an incident overwhelms the local and county resources, state officials may integrate themselves into a unified command structure for purposes of incident mitigation. The state OEM has access to a larger pool of resources than does its county counterpart. Eventually, the state OEM may make a request to the governor for activation of National Guard resources, depending upon the scope of the incident. The state OEM will also oversee its own EOC and will have an EOP. Many state OEM professionals are full-time staff members, and they too will have a line budget, which will be a part of the state budget.

Federal OEM and Resources

The last step in the OEM hierarchy will be the federal assets, which may be called upon to assist state, county and local OEM. FEMA is the largest of the federal entities that are designated to provide assistance to the requesting state agencies.

FEMA gained national exposure as a result of hurricane Katrina, which devastated Louisiana and other neighboring states in 2005. FEMA has the backing of the federal government in providing requests for assistance and resources in times of declared states of emergency. It will work with other federal agencies to coordinate the movement of manpower, tools, equipment and supplies to affected areas. These assets are moved through a very intricate system of requests and approvals. The federal system includes regional resources and clearinghouses designed to process requests for assistance from governors and provide the requested resources in a controlled yet manageable fashion.

Basic Training for OEM Personnel

All OEM staff are trained in certain areas and have a prescribed set of skills. These personnel are well versed in the administrative aspects of their respective offices. For example, staff members are responsible for establishing memoranda of understanding (MOU) and memoranda of agreement (MOA). Basically, these documents are agreements that provide for equipment, services and materials that will be provided during times of crisis. These documents are the backbones of the *mutual aid* system, which mandates that agencies or entities will provide whatever resources are requested by another entity to try to resolve an emergency. Generally, these mutual aid agreements allow police, fire and EMS assets to be requested and used across jurisdictional boundaries. The most common example of a mutual aid agreement is found in the fire service. If a town has a structure fire and needs additional apparatus or manpower, a fire coordinator or the IC will make a request to a neighboring jurisdiction to provide the requested aid. The requested resource will be dispatched to report to the scene or a staging area, as requested by the incident manager. Many times, these mutual aid agreements are very informal, meaning that there is no written agreement. Emergency service responders will assist each other any time they are requested. Automatic aid agreements, as defined earlier in this book, are a different type of aid. The automatic aid agreements provide that resources will be sent immediately upon a call for service. Mutual aid requests usually are placed from the scene, when a need for resources is identified. Some mutual aid and some automatic aid agreements are formal and, therefore written, and signed by representatives from both jurisdictions. Also, the OEM staff has certain administrative duties, such as maintenance and updating information and annexes on a regular basis, and other mandated paperwork. This paperwork is usually requested by the OEM coordinator at the level directly above the OEM coordinator to whom the request was made.

In this chapter we have identified the various levels of the OEM system. I have explained the duties, responsibilities and functions of the various offices of emergency management. I have shown that each level of OEM is designed to support the subordinate level. Also, it is imperative to note that the OEM staff are available to support the various incident managers during any incident. From this point forward, the additional training that I will argue needs to be provided to these staff workers will be addressed. Each area of training will be mentioned and applies from the local through to the federal system.

Training Proposals for OEM

Why do we need to even bother providing training to OEM staff, who will not be directly involved in the tactical aspects of incident management, and mitigation? Well, the answer is simple; these personnel need to be aware of the issues we face, the problems we will face, and the special resources and assets that will be needed during the mitigation phase of any incident or event. They need to have a shared vision of the incident with the incident managers and to be prepared to request the special equipment, tools or resources. For example, if there were an incident involving chemical warfare agents (CWA) and the IC needed a device that could monitor the environment for additional possible CWA that might be released, would the OEM coordinator understand what the IC needs, and know where to get one? This is one example, and many others can be cited. OEM responders and support personnel will play a vital role in the management of WMD incidents, MCI or other large-scale incidents or events. Since they will be heavily relied upon to bring the needed resources to bear on the incident, they must be provided ample training to complete their mission.

Nationally, all large-scale incidents will be managed under a recognized incident management system, generally referred to as the incident command system (Figure 6.1). Without explaining the entire ICS, I will relate why the OEM personnel need training in this area. The ICS has eight key leadership positions. The primary and only necessary position at *any* incident is the IC. Under the IC, there are two staffs, general and command. The command staff includes the incident safety officer, the liaison officer, and the public information officer. Each of these officers may have *assistants*. The general staff consists of the planning section chief, the operations section chief, the finance/administration section chief, and the logistics section chief. The IC has overall responsibility for the incident. The planning chief is responsible for the written plan on how to mitigate the incident, and for the units under this section. The operations chief is tasked with using the plan written by the planning chief and the resources requested by the logistics chief, and making and carrying out a tactical operating plan to resolve the incident. The finance/administration chief is responsible for tracking cost expenditures and administrative aspects of the incident, including tracking and documenting injuries at the incident. The logistics section chief is charged with reviewing the plan developed by the planning chief and making arrangements to have all necessary resources on scene for the duration of the shift of operations, as well as arranging for and staffing the facilities needed for each incident. Basically, he or she will

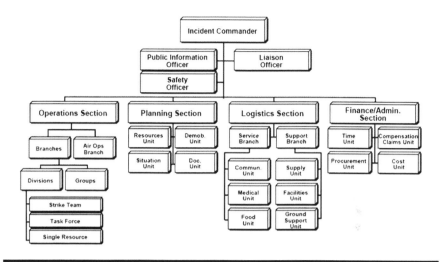

Figure 6.1 An overview of the Incident Command System, which is the preferred incident management system in use in America today. It is used to deal with large forest fires and smaller events such as local parades. It is capable of expansion and contraction, and this allows the incident managers to have only the resources needed on hand. Fire personnel use this system on a daily basis. Police departments sometimes use this management system as well. All OEM personnel should be extremely familiar with the principles of ICS. NIMS directs that all incidents be managed using this system. (Table design courtesy of Pradeep Mathew.)

request all materials, tools, equipment, manpower, apparatus, facilities and supplies that will be needed.

In the ICS, the work day is broken down into *operational periods*. An operational period, or shift, is the amount of time a responder is subject to work at an incident. On larger, more involved incidents, which may extend to several operational periods or even many days, the operational period is usually set for twelve hours. On some incidents, it may expand to eighteen hours. However, an operational period cannot exceed twenty-four hours.

Roles for OEM Personnel during the Incident

In my experience, the best, most qualified organization from which to fill the role of logistics chief is the OEM. These professionals have access to the annexes that will cover the type of incident that is ongoing and the plans needed to address the scenario. They will be familiar with the resources available to them within their jurisdictional authority, and how to order them and deliver them for use as needed. Further, they know to request assistance from the next level of OEM. Since we may be subject to very large incidents or events that may take days or weeks to mitigate, the OEM staff needs to be trained in, and intimately familiar with the protocols of incident command. Most likely, the OEM will have a major role either in being the logistics chief or in staffing one of the unit leader positions under the logistics chief's control. If we expect these responders to be used in that capacity, we need them to be well versed in the system that will be employed to resolve the crisis.

Incident command training involves classes of varying length in which small incidents are initially presented, and, at each level, the incident complexity expands, class by class. The levels are 100, 200, 300 and 400. These classes provide overviews on the *overall* operations considered at each level of training. It is proposed that OEM staff be provided specialized training within the ICS. For example, in New Jersey, the traditional classes 100 through 400 are offered. In addition to these core classes, the New Jersey Division of Fire Safety, in conjunction with the New Jersey Forest Fire Service, offers specialty classes. Examples of these include Staging Area Manager, Resource Unit Leader, and Strike Team Leader. Some of these classes are now prerequisites for instructors to be able to present classes at the 300 and 400 levels. In my experience as an ICS instructor, students are capable of "running the incident," but they do not understand each of the subordinate positions in the system. In other words, a successful I-300 student may be able to command an expanding incident and understand how to manage the command and general staffs, but may have no idea how to be a unit leader.

Everyone who is in a position of authority should be familiar with the duties and responsibilities of all subordinate positions. Therefore, OEM personnel should be given ICS training, but it should be expanded to include the specialized training to the levels to which they may be assigned. The Resource Unit Leader class would be a good start, as would be Staging Area Manager. OEM personnel could easily be in charge of any of the six units under the control of the logistics unit. Therefore, we should ensure that we train these responders who may be asked to work within the parameters of the ICS. We need to give them the training and tools to succeed. This training is very important at the local level, where most incidents are initially handled. In smaller municipalities, the local OEM coordinator will be present at the incident, and will most likely be utilized. Once the county coordinator arrives (if requested or mandated), he or she can assume the position of logistics section chief. If this occurs, the local OEM coordinator will most likely be reassigned to a subordinate position within the ICS framework. Therefore, it is apparent that this level and type of training also needs to be provided to the county coordinators and their staff.

The same level of training needs to be provided to state coordinators if they do not already have it. As noted, many responders have the core classes, but few are competent in the subordinate disciplines operating within the system. At the federal level, this type of training is not necessarily warranted. Under the ICS or IMS, the need for trained and qualified response teams has already been recognized and addressed. The federal government maintains a constant state of readiness regarding IMS response teams. There is, at all times, an incident management team on call. These teams are available, after the proper requests are made, to report to any large-scale incident in America. These incidents are detailed in the *Stafford Act,* and are often referred to as *Incidents of National Significance.* These teams will report to a location, and are prepared to integrate into the current incident management structure. The requirements and experience needed to qualify for these positions are very stringent, and take years of practical experience to achieve. Therefore, at the federal level, the training advocated for OEM at the local, county and state level is already provided.

The next area we will discuss involves the various CBRNE responses that responders may be called upon to mitigate. During a chemical attack using either nerve agents or vesicants, incident management personnel will face many challenges. If the event is one covering multiple locations that occur simultaneously, the management challenges will be magnified and increasingly difficult to meet. Already, we can concede that we will be in need of transport-capable BLS units, medical responders, decontamination teams and hazardous materials teams. Additionally, the incident managers may have requests for specialized tools and equipment. As already noted, there are many different tools available today to monitor for nerve agents and blister agents at an incident. When an incident manager makes a request for one of these tools, the logistics chief is the person who will ultimately be held responsible for its delivery. Having trained the OEM personnel in CBRNE response, they will have been exposed to some of the detection technologies available, and will not be at a loss to request the proper tool. In the annexes within the EOP, there may be equipment referenced, or the hazmat team may be able to provide this resource. However, if there are multiple sites, there will be a great need to monitor the quality of the air. Even at one incident site, if the area is large enough more than one tool may be required to provide ample real-time monitoring of the atmosphere. Accordingly, the chemical aspect of CBRNE response is clearly one example where training in advance of an incident will make the OEM responders more capable of providing timely resource requests and delivery of same.

With relation to biological incidents, OEM personnel also may be required to request certain assets and their familiarity with them will be of great benefit. Many of the biological weapons currently used will not be readily identifiable in the field. Hazardous materials teams can "field test" a substance, but may get only a positive or negative result for the presence of biological agents. If a positive result is achieved, the substance will be taken to a laboratory for further testing and identification. If there are no readily available hazmat teams, other resources may be requested. Additionally, decontamination resources will most likely be needed. Having the OEM personnel familiar with what tools and equipment will be needed will be of great assistance. The same thought process will apply for radiological and explosive incidents. Having the OEM personnel actually see the tools and equipment that responders will use can assist them when they are making requests for them from incident managers. This training is desirable at the local and county level. State OEM personnel may have it already, and if not, they too should be given the classes. Federal-level OEM will probably defer to state-level OEM for these types of equipment and tools, but even they should have the basic training.

Why Ask OEM to Train with First Responders?

Training the OEM personnel on the capabilities and limitations of our responders will make their job much easier. They need to be aware of what the responders can, and more importantly, what they cannot perform in the field. Allowing or requesting OEM staff to participate at drills and training programs will enhance their understanding of protocols and procedures in place for the different response communities. Most likely, the OEM staff will man the logistics section during an incident. This entails the acquiring of resources that the incident managers will need in order to mitigate the emergency. Additionally, OEM staffs are charged with maintaining response annexes on the EOP. These revisions can be more readily prepared if the staff conducting the revisions are familiar with tactics and special equipment that may be needed. In this manner, they can begin to look for private vendors who will be able to provide the necessary equipment in times of emergency. OEM staff at all four levels can certainly benefit from this additional training. In cases where the OEM staffs are full-time, or paid members of the agency, training is very easily completed. It will be considerably more difficult to arrange for training of volunteer staff, but the training can be provided. Understanding what will be involved as responders and incident managers are trying to control and resolve an emergency will be invaluable as OEM staff members begin to make arrangements for resources and materials during the incident. OEM staff, when used as logistics section staff will be an integral part of a successful resolution of an incident. Their ability to anticipate the needs of incident managers, as well as properly amend annexes, is of paramount importance. Knowing where to get the resources requested by incident managers is extremely important.

When resources are called for, the plans to acquire and deliver them need to be in place, and they need to have been tested in advance of the emergency. Including these responders in our training process will serve us all in cases of emergency. This scenario, where the OEM personnel have been exposed to training, will make their jobs much more easily accomplished during the crisis. They will be able to relate the tools and equipment needed from the training, and, in this case, seeing the need for certain equipment and tools will make them a far better trained asset for incident managers.

At all levels of OEM, from local or small entities through our federal management system, additional training should be mandated. Preparation is key to the rapid and safe resolution of any emergency. The time for preparation is right now. We cannot learn on the job while people lie sick, injured and dying. We have been exposed to many possible terrorist attack scenarios, and we need to have the vision to realize that they can happen at any time and at any place. We can no longer turn a blind eye and hope that the incident occurs elsewhere. We need to make the most of the opportunities we have been given to prepare our communities for the unthinkable. While we are waiting and hoping nothing will happen, there are those relying on us to have this response ethic. Spending some time, effort and money now will hopefully save lives if we need to respond to an MCI event of any type.

Finally, the OEM responders need to have frequent interaction with the other members in their "chain of command." Local OEM personnel need to be in regular contact with county personnel. The county must maintain close contact with the state OEM, and the state OEM need to be intimately familiar with the tedious procedures for requesting and receiving federal assets. Most OEM are very familiar with those in the next level and do maintain excellent working relationships. The federal assets will require time to be ordered and delivered to the incident. Regular training in this area, as well as the areas referenced earlier in this chapter will make us as prepared as we can possibly be for any incident that occurs. OEM is the department or agency that will get the responders the supplies they need, and accordingly, it needs to be as prepared and well trained as the incident managers and responders.

Chapter 7

Public Health Professionals

Introduction

The American Public Health system is a very large multilevel system, similar to the OEM system currently in place. There are local, county, state and federal components of the overall system. The same structures are in place in this realm as in the OEM arena. Small towns and jurisdictions have very small, usually part-time inspectors, while larger cities will have career personnel and larger budgets. County personnel are generally full-time career personnel, as is the case with state personnel. Federal health agencies are varied, and are staffed by full-time career personnel. The jobs are similar at all levels, with the aim of keeping American citizens as safe and free from disease as possible. In general terms, there will be much less cross-training required of these professionals than of others. The largest area of concern for public health professionals will occur in the CBRNE arena, most notably in the biological attack area. As noted in other chapters, there are UASI regions in America that receive additional federal grant funds for various projects. The public health agencies have also been eligible for these grants, and have strengthened their cache of equipment and supplies. They have been able to use funds for planning and response to emergencies. Health agencies may have used grant funds to educate citizens, purchase vehicles or even to conduct tabletop or hands-on drills to test their annexes.

Public Health Responsibilities

Public Health officials have many responsibilities in their regular course of business. They are required to make inspections of various types of public facilities, including areas where food is served, and medical facilities. Additionally, health departments will monitor illnesses reported at hospitals, being ever vigilant for spikes in certain types of illness that may indicate an epidemic, public health emergency or terrorist attack. Epidemiologists are charged with studying diseases, and will track trends for their agencies. The role of public health agencies has changed since 9/11. Americans and the response community are more cognizant of the potential for biological attacks from terrorists. This was evident in Florida and New Jersey in the aftermath of 9/11. In both of those states, there were biological attacks that used anthrax as the weapon of choice. There were several fatalities, and in New Jersey, the post office through which the affected letters were routed was closed for several years.[1] There have been other incidents of terrorism, mostly domestic, where public health officials were called in to assist with cause and origin determinations.

When there are sudden increases or spikes in certain types of emergency room visits, hospital officials notify local health officials. These responders begin an investigation into the likely causes of the spike. In most instances, there is no evil or diabolical incident or event, just a surge in a particular type of illness. However, there are times when this is not the case. In 1984 in Oregon, there was a biological incident. An individual named Bhagwan Shree Rajneeshee sent members of his cult to spray salmonella onto the salad bars in many restaurants. While there were no deaths, approximately 750 people sought medical attention.[2] This is an example of domestic terrorism, and the health department and law enforcement officials investigated the cause of the sudden influx of cases of stomach distress. The cause of the increase in cases was not immediately known. Health officials looked for commonality among the victims. In Oregon, the affected restaurants were using many of the same distributors, and this added to the delay in identifying the cause. By questioning victims, studying common eating destinations and food choices, the scope of the investigation narrowed. Eventually, the cause was determined, and after that, those responsible were identified.

Specific Incidents and Scenarios

In cases like the one in Oregon where there is a spike in emergency room visits for certain illnesses, there are two basic thought processes. One is that there could be a *public health emergency*. During these events, investigators have no initial idea why there is a spike. They must begin the intensive process of questioning victims and looking for the commonality that will help to isolate the cause. The next type of crisis is a *focused response*. In these cases, the location becomes immediately known, and the scope of the investigation is tightened. For example, if many patrons of a local restaurant all experience the same symptoms, the location of the incident is now known. The response by health officials is focused to the one location. Determining a root cause in this case will be much easier than in a public health emergency.

In the post 9/11 environment, we have seen a rise in the attention given to potential biological attack scenarios. Those who specialize in each field of CBRNE response will argue that one of these scenarios will take place and have devastating results. Chemical analysts will argue that a chemical attack scenario is most probable, and biological specialists will predict a biological scenario; explosive experts, nuclear analysts and radiological experts will also predict an attack in their area of expertise. Short of a nuclear device being deployed and successfully detonating, biological attack, in my opinion, is the most likely scenario we will face next. It has been widely rumored that weaponized biological agents are available for sale on the black market. While there is no direct evidence, no one can definitively say that this is not the case, or if they even exist for sale. If the aim of our response community is to be proactive, we would then operate under the opinion that the agents still do exist and may be available for sale. Should this be the case, biological weapons provide a significant return on the investment by the terrorist group. The relative cost to produce these agents is very low, and the loss of life, if the weapons are properly manufactured, weaponized and deployed, would be significant. For example, smallpox has a mortality rate of approximately 30 percent. This virus is very contagious. Should we be faced with a nationwide outbreak of this deadly virus, the death toll could approach one hundred million citizens. This country has already seen limited biological attacks, by domestic terrorists and unknown persons, with low death tolls. Introduction of a more potent, contagious viral agent could have devastating effects, including psychological impact. We have identified a potential problem, now we must attach the appropriate level of resources that must be utilized to prepare us for this potential event.

Response Strategies for Public Health

In response to the potential of biological attacks, the United States has begun to develop strategies to prevent or to be able to respond to any potential biological incidents. The ability to respond to these incidents has included research and development of vaccines and prophylaxis for several of the most commonly weaponized biological agents. In the event of an outbreak, public health agencies will need to be quick to respond and provide medication for the victims affected, as well as medication for those not yet affected by the agent. First responders will need to have access to prophylactic medication in order to protect themselves while they are required to be on duty. Initiatives are in place, or are pending, that will account for the need to protect our first responders in cases of biological incidents. States have been directed to develop and drill the plans for distribution of medications for these responders. In my county, for example, we have developed and actually tested our response plan.

The development of response plans addressing biological attacks include not only the response community answering additional calls for service, but also providing for a mechanism to deliver medications to those in need. Additionally, in cases of biological attack, there will be many questions from the public regarding symptoms, exposure, medical countermeasures and even questions regarding pets and their risks and medical options. Many states, if not all, have addressed most of these issues in their respective public health annexes. The ability to have the plan in place is of great value in times of crisis. Drilling, or exercising the plan, on even a municipal level, can be very difficult and time consuming.

In my county, I am affiliated with four separate and distinct response agencies. I work full-time with the sheriff's department, am an active member of the local fire department, serve as deputy OEM coordinator, and work part-time at the county fire academy. Each of these four entities has a medication distribution plan to be implemented in cases of biological attack. The various plans call for arranging for the retrieval of medication from a secure location, accounting for it, and delivering it to the first-response community. Once the responders are given medications for themselves and for their immediate families, they can then begin the process of delivery to the general public. Responders need access to the medications first, so that they can have peace of mind while dealing with the public, many of whom may be affected by the biological agent used in the attack. I have participated in these exercises on several occasions. The plans call for strict security measures protecting the medications. Further, very detailed and closely followed procedures are in place prior to agencies' being able to retrieve their allotment of medications. Lists of responders and family members are routinely updated to ensure that they reflect the current makeup of each responder agency. Due to the sensitivity of the procedures, I cannot discuss exactly what is in place regarding protective measures.

In another drill, called Top Officials (TOPOFF), my home county participated in a real-world exercise of our county plan in regard to a simulated terrorist attack. We simulated a countywide biological attack, while other areas of the state had other issues to deal with. We were required to exercise our annex for the attack, including establishing, and operating a point of dispensing (POD). To store our *simulated* medication, we chose a secure location that is not the location chosen for a real-world incident. We formulated a site-specific security protocol, and had personnel in place to test our procedures during the drill. We simulated security at our POD, and arranged for actual civilians to participate in the drill, serving as either victims or healthy people in need of medications. Drill planners, which included public health officials, factored in a number of possible scenarios that tested the abilities of those in charge to change or amend the plan on the fly. Some of these included language barriers, people who could not be issued medications for medical reasons and sick calls on the part of responders from all emergency response disciplines. State and federal leaders attended the drill, and federal reviewers assessed the drill planners. After the drill, a review was conducted and any changes required to be made in the annex were addressed. The drill was a success, and we gleaned a tremendous amount of information that was used in our corrective action plan. Since this was a federally mandated exercise, the planners were given a budget set by the federal overseers. All costs were documented, and we stayed within the established parameters set forth by the federal overseers for spending allotments for this drill. A tremendous amount of simulation was put in place to reduce costs of staffing all of the positions, which also reduced the amount of money used to pay for overtime costs incurred by participating agencies. Our drill was multistate and multinational. Our county public health department was the lead agency, and it did an excellent job. Our drill included two foreign nations as well as another state, all of whom tested part of their response plan at the same time.

With regard to the statewide drill we just discussed, there are other areas where cross-training would be beneficial for public health officials. Law enforcement was heavily involved in the site protection and security plans. However, the health officials are neither fully trained nor offered refresher training in other CBRNE disciplines, and this needs to be addressed. They should, at all levels, be provided some basic terrorism awareness, and terrorism response training. Again, if we will call them into the field to assist, they need to go in with eyes wide open and having an idea of what to expect or look for. Tendencies of suicide bombers, placement of IED, secondary or tertiary attacks, all need to be addressed. Also, since the incident managers will be using the incident command system, we must train the public health officials in that system.

Incident Management Training for Public Health

With regard to incident management training, I have already addressed the issue of training personnel in the command components of managing an incident. People with special training, such as public health officials, are referred to as *technical specialists* within the realm of incident management. These technical specialists are often used under the planning section chief, where their specialized training may be put to its best use. Accordingly, the public health community should be well versed in incident management, primarily in regard to the input planning section personnel will require while formulating the plans to be used for the incident. The logistics section will also use the expertise they have when they are ordering resources to be used by those who are operating in the field. Additionally, public health officials can be used as part of the situation unit, also a unit under the domain of the planning section chief. In this capacity, the public health official can provide real-time feedback to the planning chief, and eventually the incident commander, to allow for the adjustments and modifications that will need to be made to the existing plan. Training the public health community as to how their input can be best used in the incident management system would certainly be beneficial at an incident. On the federal level, regarding technical specialists, there are many different options for incident managers. The Centers for Disease Control in Georgia would be a great asset in the event of a nationwide or even countywide biological attack or outbreak. This federal agency is one of the best resources for identification and treatment options for disease in the world.

Should this country be subjected to a biological attack or any type of national outbreak of a virus, our public health officials will be ready to respond. In the post 9/11 era, legislators on the county, state and federal level have seen the need to provide financial support for public health officials. The threat alone of a biological incident was enough to prompt a quick review of the equipment and procedures currently in place at all levels of public health. In most levels of public health response, policies were revised, personnel were exposed to new, more modern equipment and they were given additional training. The overwhelming amount of this training was geared toward refreshing existing training the responders already had and possibly some new training on some of the biological weapons that could possibly be used against us in an attack. While this was necessary training, and I readily concede that this was the proper group to receive this training, we can do better. Public health officials will also be called upon to provide expertise in the areas of chemical attack, and with radiological attacks as well. The bulk of the public health community, which will probably be on-scene, and operating in the field, will be local and county officials. These responders will be required to be present to distribute medication to those exposed to whatever agent is used and to provide doses to those not yet affected, but required to report for work. Providing these responders with additional training in other areas will allow for a smoother operation and give incident managers additional and better-trained assets that can be used in the resolution of the incident.

Endnotes

1. Newark (NJ) Star Ledger. "A timeline of the anthrax case." August 1, 2008. http://blog. nj.com/ledgerarchives/2008/08/a_timeline_of_the _anthrax_case.html. 8-1-08.
2. FEMA-US Fire Administration-National Fire Academy Course Offering entitled "Emergency Response to Terrorism: Basic Concepts." Student manual pages 4-3 and 4-4.

Chapter 8

Other Responders

Introduction

Other groups of people in America may be present when disaster strikes or may be able to help to prevent the disaster altogether. The first group to review includes many different types of workers, some of whom one would expect to be able to assist in prevention and response, and others who one may not even realize can be of service. This group of citizens is very large and represents every state and county in America. They are what we can refer to as our "responders hidden in plain sight." This group represents many workers whom we see on a daily basis and generally do not even notice. Others are not seen as often, but we do not think of them as possible responders or preventers. In fact, if we were to ask many of the people we will include in this group, most may not even realize the great assets they can be to us in many different areas regarding emergency response. Throughout the remainder of this chapter, we will discuss who these potential responders and preventers are, what types of training they should receive, where to get it and how this can be paid for.

Mall Security Personnel

The men and women who patrol our malls will be our next area of focus. Every state in America has malls in one form or another, many of which are enclosed. Many of these larger enclosed malls provide shoppers a variety of shopping options within one large area (Figure 8.1). These malls are usually made up of one or more cornerstone stores with regional recognition, plus many smaller specialty retail stores.

Figure 8.1 Enclosed malls, some consisting of two or more levels, are attractive to terrorist groups. At peak times such as the holiday season, these malls are packed with shoppers and the parking areas are filled to near capacity. The weak security controls at mall entrance areas make these facilities very desirable targets for terrorist groups to plan and carry out an attack. Additionally, should an explosive attack be planned as a secondary attack, the parking lots are an excellent target area of opportunity.

Food courts and even movie theaters also support the mall operations on a daily basis at most of these mega malls. One common thread among these large malls is that they all provide a small security force, which in turn provides support for law enforcement. These security personnel generally have no police powers, and will notify law enforcement to assist them in the event of criminal activity. They provide services to the mall visitors such as directions, lost child assistance and perimeter patrol to protect vehicles from intrusion and theft. They also usually provide the security force that will monitor all of the security cameras in place in the common areas of the mall, and are trained in loss prevention techniques and shoplifting trends. Some security personnel have vehicles to provide these services, but most are on walking patrol of the mall areas to provide a visual physical deterrent to anyone who may have criminal acts on their minds. In my experience, the demographic for mall security guards is either people retired from other careers who are desirous of additional income, or very young people wishing for law enforcement careers. The majority of these mall security personnel are males, but there are females as well.

Mall security personnel are not very well trained outside the areas that mall management would like them to be. They are generally affable, gregarious people who do a good job. To succinctly characterize some of what they are trained to do, is that they are *people watchers*. If we are looking for trained personnel to observe strange or criminal behavior, then the mall security force is an excellent tool for us. They scan parking lots, common mall areas and security cameras in search of unusual or suspicious behavior. These working professionals may become first responders should there ever be an attack on any one of our malls. It does not have to be the Mall of America, located in Minnesota. Large shopping malls are very prevalent all across America. Domestic as well as international terrorist groups have long seen the value for their groups in targeting shopping malls in America. The psychological effects alone of such an attack would be devastating. Given the proper date and time, a carefully planned and executed attack at any of our malls would produce a large number of casualties and injuries. Terrorist planners from all security fields in America have been trying to harden, or lessen, the attractiveness of such targets in America for years. Sadly, there is only so much that can be done and *soft targets*, such as malls, casinos, schools and similar occupancies will always be attractive targets for terror groups.

So, where do we begin when discussing possible training agendas for security personnel at our nation's malls? Due to their very makeup, daily activities, and large population of civilians, malls are a rich target environment for terror groups. These facilities are attractive for several types of attacks, but not all of the CBRNE scenarios are ideal for malls. The first area we will explore is a chemical attack. Whether we are discussing nerve agents such as sarin or VX gas, or vesicants such as mustard gas, enclosed malls are very suitable for attacks involving chemical warfare agents. The nerve agents and vesicants, or blister agents, are far more destructive and deadly in warmer atmospheres and enclosed areas. Factor in the ventilation and air-handling units along with a large number of possible victims, and a mall presents itself as a perfect target for these types of agents. In one of the CBRNE classes I teach, there is a scenario of a small attack on an enclosed mall, specifically, a nerve agent attack on this mall. I have seen variants of this video example in other training classes that I have attended. Using the nerve agent as a model, we will explore a possible attack scenario in an enclosed mall.

Scenario in an Enclosed Mall

Prior to planning and carrying out an actual operation, terrorist groups generally research their targets very well to include the busiest time of day, dates, locations, responder response times and capabilities and projected casualties. When properly motivated, the members in charge of the intelligence gathering, processing and interpreting the data are well armed and prepared to conduct their attack at the most opportune time for their cause. This will almost always mean that it is the worst time for the victims and responders. Accounting for good reconnaissance and intelligence gathering, the following scenario will reflect a good time for the terrorist and a bad time for victims and responders. The name and location of the mall are not important, but the mall chosen is an enclosed mall with between fifty and sixty-five individual stores contained therein.

The day is the Friday after Thanksgiving, often referred to as *Black Friday.* The mall has a generous-sized food court, and the attack will commence at approximately noon, when the food court is at its busiest. This mall has an extremely large seating area for the patrons to sit and eat their meals and the entire food court area is of open design. The mall is located in a jurisdiction that provides volunteer fire and EMS services and has eight patrol officers on duty during the day shift. Mall owners have planned for the tremendous volume of shoppers anticipated to come to the mall for the seasonal bargains, and has added an extra six security personnel, bringing the total for today to fourteen. The parking areas are full, and the mall is close to its maximum capacity. Individual shopkeepers also anticipate additional shoppers, so they too have decided to pay for extra staff at their locations. Shortly after noon, near the center of the large seating area, several shoppers begin to experience eye pain, severe coughing, running noses and difficulty breathing. The nearest security guard notifies his control center and advances to the area for closer inspection. Additionally, other shoppers close in on the area to see what is going on and to possibly offer assistance. As the security guard nears, he too falls victim to the agent that has been secretly deployed. Most of the people in the food court realize that something is wrong, and those not yet affected begin screaming and running away. At this point, one hundred total victims are incapacitated and in need of medical aid. Additional security personnel are quickly advancing toward the food court. They are getting little information from the control center, other than it seems as if there are many people affected by something. Security staff personnel now on location are victims, and are unable to provide any real information to their colleagues who are racing to the area. At this point, control calls the 911 center to report multiple sick people and people down. The 911 center dispatches police, fire and EMS to the scene. There is mass confusion throughout the mall, and shoppers are exiting the mall as fast as they can. The sudden unexpected mass exodus has now had serious effects on the ability of responders to get to the scene. Other than the victims, the food court is now completely empty of shoppers and security officers. Of the fourteen security staff on duty, five are still in the food court, and four others were exposed briefly to the agent and were able to self-evacuate, but are in need of medical treatment. Control, located on the opposite end of the mall has not yet been affected by the agent, and is advising the 911 operator of what observations he is making regarding the scene. He now reports that all of the victims are on the ground, suffering from what appears to be seizures. The mall shoppers are still trying to exit the mall, and police officers are now arriving.

Figure 8.2 Security staff at malls may be the first to respond to any incident there. While some police departments may opt to have a substation at malls, the security officer is still likely to be first arriving at an incident. These non-traditional first responders should have access to better, more wide-ranging training.

The purpose of the scenario presented here is solely to demonstrate a potential attack at a shopping mall anywhere in America. The scenario depicted the best possible factors for the terrorists, and was designed to exploit the worst possible factors for responders. The events described have never occurred in America, but similar events have occurred in other areas of the world, where this scenario may have had different outcomes. The American response community, as noted, has tried to become more proactive in the post 9/11 era. We have spent copious amounts of money on new technologies and equipment (Figure 8.2). Newly trained response teams have been formed in areas perceived to be of high target value to the terrorists. We have evaluated our schools and malls, as well as other soft targets, but have done very little to actually make them less attractive targets for terror groups. Again, we have identified a weakness, and now, before something happens, is the time to address the problem.

In this case, CBRNE training for the security personnel would have been beneficial. The staff would have been aware that malls present certain opportunities for terrorist groups, and they may have had heightened situational awareness.

Recognition of signs and symptoms of exposure to nerve agents would also have potentially had significant impact on the incident. Other security staff could have begun to evacuate the area rather than rushing in, thus becoming victims themselves. Further, no steps were taken to secure or turn off the air handling equipment at the mall. Potentially, the agent could have been circulated throughout the mall, and the number of victims could have been greatly magnified. Awareness of attack methodologies and likely target areas within the mall could have also helped here. We do not generally see mall security officers as first responders, but they most certainly are. If they are going to be first on the scene, we need to provide them with training to recognize what is going on around them, and train them to become the eyes of incoming first responder units. They can then provide excellent on-scene intelligence for other responders who are en route to assist them.

Explosive Attacks at Malls

The last area of response for mall security staff we will discuss is an explosive attack scenario. I will also include armed encounters with gunmen who see malls as an attractive venue for their aggression. In America, we have seen shootings at malls, and, sadly they are not as infrequent as we would like them to be. However, we have not yet seen explosive attacks at our malls, including IED and VBIED or suicide bombers. Since many mall security personnel are unarmed, we will focus on explosive attacks. Armed security personnel can respond to attacks with firearms, and will undoubtedly request police response as backup. Additionally, most citizens have the survival skills needed to survive an armed encounter, specifically to seek cover and escape.

Explosive attacks will have devastating consequences if used against our society at an enclosed mall. Many people will become victims; some will be killed while others will be seriously injured. To make matters worse, there are very few techniques we can use to prevent explosive, or even worse, suicide bombings in malls. Most of the patrons at malls will have packages, and people carrying bags are a normal event at most malls. In inclement weather, bulky outer garments will probably be worn by patrons, providing another potential hiding place.

Properly planned and executed attacks of this type at malls have rarely happened, so there has been no pressing need to deal with them. This type of thinking can no longer be accepted. Simply because there has not been explosive incidents in our malls is no reason to believe there will be none in the future. Small explosive devices can produce significant damage, and pipe bombs and suicide belts would be easily secreted into most malls. The suicide bombers are truly the best *smart bombs* in the world. Injuries would occur to those close to the pipe or suicide bomber, most resulting from the blast wave and shrapnel. The blast wave will also break windows inside the mall, which will cause multiple injuries from flying glass. Victims of the blast wave will be seen for hundreds of feet in all directions, as the blast wave will be 360 degrees. In enclosed malls, the blast wave effects will be magnified, causing more injuries. Similarly, awareness about vehicle bombs would be desirable, since the mall offers a tremendous number of vehicles and parking areas, affording the terrorists many options for placement of a car bomb. If a terror group were to detonate several pipe bombs, shoppers would run for the exits, and the perceived safety of the parking lots and their vehicles. This could possibly be a case of leading the lambs to slaughter. An illegally parked vehicle loaded with explosives near an exit detonated remotely by the terrorists once the area is filled with fleeing shoppers would also produce significant numbers of casualties and injuries. The car bomb would be the primary device, with the smaller devices being placed in other areas and used as a stimulus to get victims to flee to the area of the primary bombs.

Additional Training for Mall Security Personnel

Here again, there is an identified need for some additional training for security personnel at the malls. Basic awareness training would suffice in this area, since we have conceded the ease with which terror groups can deliver explosives to the target area. If a terror group is going to deploy an explosive device at a mall, they could possibly use the bomb as a dirty bomb, or RDD. They could also use the bomb as a *puff event*. A puff event is when someone observes a small charge used to disperse chemical agents, biological agents or radiological contaminants. The small amount of explosives used produces a small white cloud or a "puff" of smoke. When someone sees the puff, they are drawn to investigate what happened, and generally can become a victim. A bomb used for any of these scenarios is still an explosive incident. Providing the security personnel with some training in this area may provide dividends in the event of an actual bombing event in our malls.

Now we are left to discuss what specific training these responders will be given, and how it will be paid for. Security staffs are paid, and their basic and in-service training is paid for as well. Private employers must be shown the need to provide this as part of the basic pre-employment training for mall security staff. They should be given training in the area of explosive devices and chemical attack scenarios. They will need to be familiar with all forms of chemical attacks, including irritants such as pepper spray. Nerve and blister agents will be reasonably effective in enclosed malls, and therefore training is needed in this area as well. Radiological awareness training may have limited value, but since the staff will have no means of detection, it can be covered under explosive training. Investing a little more time in training and preparing these responders for cases that may develop is a good step in protecting our citizens at malls. Failure to address the problem now may have dire consequences down the road. Proactivity demands that we address these training options for mall security personnel. The investment of time for this type of awareness training is very small, but the returns can be large. These responders need only some basic instruction, which can be given by law enforcement, fire and hazardous materials officials.

School Staff

The next groups of potential responders in need of training are the teachers and staff at America's schools. Unfortunately, there have been a number of attacks on schools worldwide, and we in America have seen these attack scenarios first hand. There was the attack carried out in Columbine, Colorado, on April 20, 1999. In that case, two students who attended the high school used firearms and explosives in a deadly attack that gained national exposure. This attack resulted in the deaths of thirteen, mostly students, and the injuring of approximately another twenty-five. Both attackers, Dylan Klebold and Eric Harris, the gunmen, killed themselves, thus ending this standoff.[1] There was another deadly attack almost eight years later, on April 16, 2007, resulting in the deaths of more than thirty people. This attack took place on the campus of Virginia Tech University in Blacksburg, Virginia. The gunman in this case, Seung-Hui Cho, also turned his weapon on himself, ending the incident.[2]

Below is a listing of shooting incidents in schools throughout the United States of America, starting in 1999, ending in 2008. Take note of the different regions of the country that were affected, the numbers of victims and the ages of the shooters.[3]

February 2, 1996, Moses Lake, Washington. Two students and one teacher killed, one other wounded when 14-year-old Barry Loukaitis opened fire on his algebra class.

February 19, 1997, Bethel, Alaska. The principal and one student were killed, two others wounded by Evan Ramsey, age 16.

October 1, 1997, Pearl, Mississippi. Two students killed, seven others wounded by Luke Woodham, age 16, who was also accused of killing his mother. He and his friends were said to be outcasts who worshipped Satan.

December 1, 1997, West Paducah, Kentucky. Three students were killed and five others wounded by Michael Carneal, age 14. The students were killed as they sat in a prayer circle at Heath High School.

December 15, 1997, Stamps, Arkansas. Two students were wounded. Colt Todd, age 14, was hiding in the woods when he shot the students as they stood in the parking lot.

March 24, 1998, Jonesboro, Arkansas. Four students and one teacher were killed, ten others were wounded as Westside Middle School emptied during a false alarm fire drill. Mitchell Johnson, age 13, and Andrew Golden, age 11, shot at their classmates and teachers from the woods.

April 24, 1998, Edinboro, Pennsylvania. One teacher, John Gillette, was killed, and two students were wounded at a dance at James W. Parker Middle School. Andrew Wurst, age 14, was charged.

May 19, 2008, Fayetteville, Tennessee. One student was killed in the parking lot at Lincoln County High School three days before he was scheduled to graduate. The victim was dating the ex-girlfriend of his killer, 18-year-old honor student Jacob Davis.

In reviewing this disturbing listing of school violence, there is an obvious need for us to begin to better educate our teachers in certain areas. Please note, this list depicts only actual shooting incidents, and does not contemplate incidents where explosives or other weapons were brought into our schools. Law enforcement has recognized a need for better training when responding to school shootings. This awareness was amplified after the Columbine shooting, where there was tremendous debate regarding the tactics employed by initial responding police, as well as the responding SWAT teams. Many law enforcement agencies now mandate training for their personnel specifically for armed encounters with shooters in schools.

Changing the Way We Think at Schools

After studying trends in terrorism, it is painfully obvious that the old-school thought processes of targeting certain people as victims has been eliminated. This process would never target children, civilians or the elderly for acts of violence. However, in the new world of terrorism, these populations are seen as targets of choice. The violence we have mentioned deals only with incidents in American schools, and the perpetrators were not members of any terrorist groups, such as Al Qaeda. In fact, in almost all of the school shootings detailed, the gunman had some connection to the location of the shooting, if not the selection of victims. Teachers in schools today are being taught to lock themselves and their students in the classroom should a shooting incident occur. But what if a terrorist group targets us for an attack in our schools using a different attack methodology? Will our teachers be prepared for such an occurrence? We can all agree that if they are not, we should take the time to train and prepare them for such an occurrence.

Where do we begin preparing our teachers to be better able to detect and respond to an attack on our schools? The two most common attack scenarios where children could be targeted involve firearms and explosives. Therefore, we should take time to prepare the teachers and school administrators for possible incidents involving these two attack scenarios. Basic training in response to active shooters in our schools is currently ongoing, and is certainly warranted. School officials are trying to *harden* the schools themselves by adding better lighting, security cameras and even *school resource officers,* or police officers who are stationed at certain schools. Better security screening techniques and equipment would certainly minimize the risk of firearms and other weapons being brought into the schools.

A Wake-Up Call for Teachers and Administrators

The other area where our children will be at risk while at school involves explosives. Our school administrators and staff need to be aware that the potential for this scenario not only exists, but also is a very strong possibility. I attended a course of training on soft-target awareness that was held in Atlantic City, New Jersey. I attended this three-day class with a colleague, and we were among only a handful of law enforcement personnel in the class. On one of the days, the course focused on schools in America. During the course of the class, we were broken into groups of approximately eight to ten students. The activity for our groups was to plan an attack on a school, and we were not to allot too much effort into the logistics, such as how we would get our equipment needed for the attack. Each group was afforded a twenty-minute time frame to plan an attack, and then we were to report back to the instructors, who would evaluate our attack scenario and pass judgment on the plan. My group included my colleague and school staff, including teachers, a middle school principal, and school board members. Once we were tasked with our assignment, the other members of my group were obviously disturbed at the nature of our group assignment. In fact, the planning of an attack on any school was not discussed at all. The group sat in disbelief that there were people who would even think to target children while they were attending school. My colleague and I were equally amazed at the naïvete displayed by those who were in charge of protecting our children. The point here is that there are those in this world who mean to do us and our children harm. No other demographic in America will generate as much anger and passion as would an attack on our children. I am not saying that all teachers and staff at our schools are this naïve, but we cannot afford to have anyone remain not fully aware of the risks.

To plan an explosive attack on one or more of our schools would not require much advance planning for a terrorist organization with prior experience. In fact, we are often very helpful to terrorist groups who will gain intelligence from simple observation of their targets on a daily basis. For example, schools are required to conduct regular fire drills, so that, in case of an actual fire, the students will be familiar with reporting locations and instructions. Additionally, should a bomb threat be called in, the school may be evacuated, and therefore, the terrorists can see where the children will be located. Generally, the children are moved away from the building, so as to allow responders to make entry and resolve the problem. Often, students are put in parking lots during these building evacuations. Accordingly, if the terrorist either calls in a bomb threat or activates the fire alarm, or simply waits for the scheduled fire drill, they will see where administrators place the students. Once this information is at hand, the planning of the attack can occur, and if done properly, devastating damage and loss of life will occur. One or two vehicles loaded with an explosive will certainly produce a large number of casualties. And, if the terrorists are really trying to make a statement, one or more vehicle bombs can be left for the fire, police and EMS responders. Can this scenario really happen? If the terrorist group targets a school, it may very well happen.

Hardening of the schools in America is a great idea, and will certainly make parents feel more at ease to know that the welfare of their children is truly important to school administrators. The use of security cameras is also a good idea, but only if the cameras are monitored and those monitoring them are trained in what behaviors and actions need to be watched. Providing educational staff with terrorism awareness, explosive awareness and even situational awareness can serve to provide an extra layer of protection for our students. Teachers are always looking out for the well-being of their students, and they are aware of sexual predators and will be looking for lone adults lurking near school grounds. This type of behavior is a form of situational awareness, but the teachers are not fully situationally aware of terrorist activities. Properly trained and motivated terrorists will not fall into this category, and will be much harder to detect by our teachers and administrators. Teachers have many responsibilities already, but if we provide them additional training, they may be able to detect a person or group with evil intentions regarding our children.

Again, we have identified an existing area of weakness in protecting our children, and now, we must find options to correct the problems. Almost every state has time during the year when school is out but teachers must work. These days are referred to as in-service days or teachers' conference days. Regardless of what they are called, these days are an excellent opportunity for America's teachers to be trained in terrorism awareness and bomb awareness. While it is conceivable that a school is also a good target for a chemical attack, the likelihood of such a scenario is very slim. The ability to gain entry and place a device without being observed is a stumbling block for the terrorists.

Incident Management for Administrators

The last area of training for school administrators to consider will be in the area of incident management. While teachers will most likely not be involved in the management of an incident, school administrators may very well be placed in critical areas within the incident management system. They will be able to provide intelligence for responders to consider when formulating their response plans. The administrator may very easily be assigned to either a unified command position, or possibly a critical position within the intelligence unit. Therefore, to provide incident command training to the administrators of schools is certainly worth pursuing. As with other responders, having these people trained in the basics of incident management will make for a more effective integration by them into any incident management system in use at an incident. The basic areas of training for teachers and administrators can be easily delivered to the staff during mandated in-service training. The cost of this training should be nominal, and if relationships are well established between school administrators and the response community, the training may end up being free of charge to the staff. In a later chapter, free training options including incident management will be discussed, along with contact information.

School Bus Drivers

The last group of people associated with our schools who need training comprises the school bus drivers. These men and women are charged each day with the timely and safe delivery of our children to and from their schools. They must meet certain state requirements before we will allow them to drive our children anywhere.

However, while the bus drivers will probably not be seen as first responders, and will probably not be requested for sensitive duties at an incident, they are still an excellent resource for us. They are adults who, on a daily basis, operate a vehicle through many of our local surface roads, and are responsible for the safety of dozens of children. With proper training, these people can be vigilant observers for things that do not seem ordinary. Events such as adults who have no children watching the students boarding and exiting the buses, for example, should raise a red flag. The bus drivers are constantly observing traffic to be sure the children are safe to cross roads, and are checking mirrors to be certain that all children have cleared the bus prior to the bus's departing for the next stop. These people are very much situationally aware of what is going on around them. But, if we were to give them some terrorism awareness training, they might be in a position to detect the presence of potential terrorists who may be gathering intelligence for a future attack on our children. This type of training would be very brief, and the cost should be nominal, save for the hourly wage of the driver. But, taking advantage of a rolling platform with the ability to observe unusual activities and report them to law enforcement is worth the small financial investment.

Airport Personnel

The next broad area where additional training can be beneficial will be the many airports and airport staff, including flight crews. After the 2001 terror attacks, airport security came under great scrutiny, since all of the weapons used by the hijackers went through security checkpoints. Policies and procedures were revised and updated, and many items that were previously allowed past security checkpoints, but *could possibly* be used as weapons were banned. But even these revisions were not the final word. After Richard Reed tried to detonate his infamous *shoe bomb,* policies regarding the bringing of liquids on board aircraft were modified, and strict limits on the volume of fluids allowed were developed.[4] The Transportation Security Administration (TSA) has made great strides in making airports more safe and secure since 2001. There are more advanced metal detectors in use, x-ray machines have also been upgraded and there are other security technologies under development. Also, the personnel assigned to operate and monitor them are getting much better training. These workers have become much more situationally aware after the terrorist attacks, but there are always areas where we can provide better training or upgrade current systems.

Defeating Technologically Advanced Weapons

Weapons systems are becoming much more technologically advanced, and many firearms and knives are now made primarily out of composite materials similar to plastics. While we have acquired better metal detection technology, these metal detectors do not detect plastic and other composite weapons. It is very difficult for screeners to focus on the x-ray machine images and the metal detectors *and* expect them to try to scan the facial and body-language indicators that signal something is out of the ordinary. Rotating these personnel from post to post during their shifts is a good idea that has been implemented, and this should allow them to maintain a better level of situational awareness. The TSA provides a good basic course of instruction for its employees, but this information needs to be refreshed and updated on a regular basis, as for all other responders. Terrorism awareness is a good place to start in-service refresher training. Further, suicide bomber awareness training will be a good addition as well.

In most airports, patrons have at least some access into the building before they are subject to security checks. In busier airports, a suicide bomber outside the purview of security checkpoints could arm his or her weapon and detonate it outside the secure area. Obviously, such a scenario will result in tremendous loss of life, as well as great numbers of additional victims. Law enforcement officers are generally assigned to patrol airports, and they should have this training prior to being assigned to the facility. Providing these two basic classes to TSA employees may put them in a better position to detect the possibility of a terrorist attack in the airport. It is conceded, however, that a suicide bombing is the most effective terrorist tool, and the most difficult to prevent by law enforcement. Given these two factors, it is imperative for the safety of all air travelers that the personnel charged with airport security receive the best available training, and that it is refreshed and maintained.

Airports and aircraft are still targets that interest some of the international terrorist groups. They realize the difficult position airport screeners are in, and they are also aware that technology has made smuggling weapons onto the aircraft somewhat easier. So, are the TSA security personnel really the last line of defense when it comes to airport and aircraft security? Well, not really, because there are two more layers of protection for travelers. The first and obvious layer is the air marshal. These professional federal employees are very well trained, and have a targeted scope of expertise regarding aircraft safety. The marshals are armed and are provided excellent training in terrorism awareness and firearm use. They are designed to be hidden in plain sight, and are chosen to be the last line of defense. The air marshals are not present on every flight, and there are not nearly enough of these professionals in the country to protect every flight. However, we can assist these professionals by providing some basic training to the flight crew.

Each passenger is required to present a boarding pass prior to boarding the aircraft and are then usually greeted at the cabin door by flight attendants. If we were to give these staff some awareness training, they could support the precious air marshals in detecting the presence of potential terrorists. This will be a delicate balance for them, because nervous travelers who have a fear of flying also exhibit many traits of terrorists. However, the ability of flight crew personnel to home in on a passenger, or group of passengers, who are displaying certain indicators or warning signs would be very beneficial to assisting the air marshal in securing the aircraft. The thought process here would be to allow them to make independent evaluations of passengers entering the aircraft. If adequate indicators are present, the staff can make a request for further evaluation of the passenger(s) who are displaying potential indicators of terrorist activity.

Sometime after the attacks in New York, Virginia and Shanksville, Pennsylvania, a survey was presented to American travelers. The overwhelming majority of those questioned indicated that they would be willing to get to the airport three hours prior to their flight if better security could be achieved. I wonder how many of those same people would answer the question the same way eight years after the attacks. They would probably not, since we are a country of people with short tempers and even shorter memories. Since the airline industry has not seen a similar attack, we feel that all is right with the world again. Having flight crew personnel observing passengers and reporting abnormalities is probably a good thing, but anxious and short-tempered passengers may make them leery of reporting their observations. However, the possible benefits of additional trained personnel observing passengers in an area where terrorists still have a genuine interest can only assist us in becoming more secure.

Private Security Staff

The next area of non-traditional, potential responders is the private security force currently working in America. Many private-sector companies hire private security guards to patrol their facilities, and protect their businesses. Some of these security guards are armed, and some are not permitted to carry firearms. Regardless of their ability to carry firearms, they are a potential source of responders, as well as a source of intelligence gathering for law enforcement. The mall security personnel have been separated from this group, and were addressed earlier in this chapter. Security companies across America provide security officers at a wide variety of locations, from warehouses to retail stores and even chemical facilities. They are present in hospitals, parking facilities, tourist areas and parks. These responders are located at what might be referred to as "target areas of opportunity" for terrorist groups. They are present to protect certain assets, although at certain targets such as parks there may even be law enforcement present. Providing these resources with even limited training in the areas of IED, terrorism awareness and suicide bomber awareness will greatly enhance their job performance with respect to observing the potential for terrorism. Terrorists may decide to seek alternative targets that are not protected by security personnel, but they may also see the presence of security officers alone as making the target even more attractive. The deterrence ability of private security is not as high as it is for law enforcement officers.

Given the scope of the areas where private security forces are prevalent, providing them with training and soliciting input from them would seem natural. As with mall security staff, most private security personnel are generally men, and are either very young, or retired and looking to supplement income. The younger guards are usually interested in law enforcement careers and therefore will work well with police and sheriff's personnel.

Private security personnel are trained to become familiar with their workplaces and the people who work there. They are present to conduct security sweeps of certain locations and are therefore familiar with the layout and features of their assigned location. Many of the facilities that contract for private security personnel are soft targets and many are attractive to terrorists. For example, any type of chemical facility that either produces, manufactures, stores or ships chemicals is a potential target for terrorists. Their employers provide the security personnel with basic training that is geared mostly toward policies and procedures of the company and loss prevention. As indicated, locations that employ private security are often of interest to terrorists and therefore additional training should be provided to those maintaining the security vigils at these locations. Since these security officers should not be intimately involved in incident management during the response, the scope of their suggested additional training will be limited.

Specific Training for Private Security

As with all of the other miscellaneous responders, terrorism awareness is a great place to start. Private security officers are the eyes and ears for the facility to which they are assigned by their employer. They become very familiar with their surroundings and, given proper guidance and training, they may be a tremendous asset in detecting the presence of terrorists. Training on surveillance techniques and locations of terrorist groups would allow them to support the efforts of law enforcement to monitor potential threats to the facility to which they are detailed. Law enforcement is very familiar with the potential target sites in their response areas, and will try to closely monitor those locations for the presence or indicators that terrorists may be nearby.

Unlike police, security personnel are on these locations around the clock every day of the year. If law enforcement has the ability to have someone monitor these locations at all times, that is something we should take advantage of. Again, we are not trying to make these guards into secret agents, but rather, given a little training, and some guidance, they can become great assets to law enforcement officials. Training them in other areas such as CBRNE would not be required at all levels. Since there is always the possibility of explosive attacks at certain facilities, basic explosive awareness would also be recommended for security forces. Simply providing these people with information on where explosives may be located and how they may be deployed could be of tremendous value to them in the event of an incident. This training can be easily implemented into basic training for the new employee during orientation, and can be provided rather quickly to existing employees. Our leaders and administrators need to push for this type of training regardless of the additional cost to the security guard companies. Yet again, we see a potential vulnerability, and there is a capable, competent possible solution already available. All we need to do is to develop the resource, and provide some basic training. If cooperation is not garnered, perhaps the state and or federal legislators can be enticed to enact legislation compelling better and more comprehensive training for security officers.

Casino Security Personnel

The next group of public employees I will discuss is a small group that falls within the private security sector. The final group is that of security personnel at casinos. Many years ago, only two states—Nevada and New Jersey—had casinos that attracted national visitation. However, more and more states are realizing the value of legalized gambling and, therefore, casinos are becoming more prevalent. The casino industry has tremendous security resources, invests a great deal of money into this area and, like mall guards, their security forces are well trained in *people watching*. This is a tremendous skill set for law enforcement to use in the war on terror.

Due to their popularity and large numbers of people, or potential victims, located inside them, casinos present an attractive target site for terrorists. Certain terrorist groups are known for their tedious surveillance and studying of targets prior to assembling an attack plan. These terrorists are also people watchers, like our security personnel, but they are watching for tendencies and security weaknesses. Additionally, casinos are open almost around the clock, and there are always potential victims for terrorists. Security developed for casinos is very technologically advanced, using facial recognition software to identify persons known to defraud casinos, or who are not permitted in casinos. The security staffs at the casinos are very in tune with indicators of possible cheating behaviors exhibited by patrons. If these security personnel can detect patterns of cheating and identify certain indicators of cheating, they could then conceivably be trained to look for other behavioral cues in people. If we were to train them in behavioral cues indicating terrorist activity or pending violence, this would be invaluable in securing a soft-target source for terrorists. The casinos have no security measures in place to prevent visitors or players from entering or wandering around the casino floor. In fact, many of the visitors will take their time in selecting the specific gaming table or slot machine to use. The lack of any security checkpoints makes the casino a desirable target for certain types of terrorist attack, most notably a suicide bombing. Ease of entry, coupled with large numbers of victims, makes the casino industry a high-value target for terrorists. Providing additional training in the areas of suicide bomber tendencies and basic terrorism awareness can provide a more secure environment for the casino patrons.

National Guard and Reservists

Finally, in the context of other responders is a very large group, probably the largest single group that falls into this category. This group is our National Guard and military reservists. In the event of an incident of national significance, after going through the proper channels, the National Guard will most likely be requested to respond to the area of the incident to provide whatever support is requested of them. The members of this group have a tremendous advantage over all of the other responders in this category. They have already received the great majority of the training that is recommended for other responders. They are not only trained, but also have some of the equipment to which other responders do not have access. These responders have had the military version of CBRNE training, and have access to APRs for use in certain types of contaminated atmospheres. These men and women are used to working in a system similar to the incident command system, which will be used to mitigate the incident. They are familiar with concepts such as unity of command and span of control, and will need no additional training in this area. The senior officers may need to be trained in the civilian version of incident command, since they may be integrated into key roles at the incident. The members of the National Guard may need only to get some refresher training in terrorism awareness and be briefed regarding their roles in responding to domestic incidents. Since we will not expect these resources to be dispatched to fire calls and calls for medical assistance, they will probably represent the group with the most cross-training. The roles of the National Guard at large-scale incidents will be mostly law enforcement in nature, and these troops are well versed in this arena. These resources will definitely be seen as force multipliers, and will allow rank-and-file law enforcement responders to handle their regular volume of service calls. Therefore, while they are put into this group, they are in need of very little training to be of great assistance to the incident manager.

Summary

This chapter has explained how certain workers in America can be utilized by law enforcement to become force multipliers with regard to possible terrorist activity detection. Shopping malls are a very attractive target for terrorists foreign and domestic, due to the number of citizens in attendance and the fact that the malls are enclosed. Mall owners employ a security force that can be used to supplement the local police response in the event of an attack. However, we must train these resources in strategy and tactics of terrorists, as well as attack methodologies, in addition to signs and symptoms of exposure to certain chemical weapons.

We have also explored the educational arena and identified areas where teachers and administrators can be exposed to additional training to make schools and children much more secure. This was followed by a review of airport and aircraft personnel, and what training may make these areas safer for travelers. Private security firms and casino security staff also provide valuable services to their respective employers. Finally, the National Guard was discussed as possible responders. Their training is very adequate, and accordingly, they have little need for training. However, it would certainly not be a bad idea to provide these men and women with refresher training, if they needed it in any particular area.

These under-utilized assets can certainly be used to better advantage in the field of detection of suspicious persons and suspicious activities. Providing the additional training these assets would require would not take too much time. There are academy staff, law enforcement, fire department and hazardous material team members who could provide the training. Delivery of the training materials can most likely be free to employers, but they would have to compensate the employees for their time to receive this training. Employers providing their clients a better trained and more highly skilled security staff can offset this investment of financial expense. The more people who are looking for indicators of terrorism, the better off we all are. The only questions which need to be answered, are does the public feel we are doing enough to protect them, and do they think that by requiring this additional training, that maybe we are going too far.

The public support for new initiatives is always tied to current events. Most Americans will support a new policy if it is in reaction to an event that has already transpired. However, there will always be two sides to this equation. After the London subway attacks in 2007, the New York City Police began random inspection of certain packages being brought onto subway cars. While the public was generally in support of the protective proactivity, the American Civil Liberties Union (ACLU) was of the opinion that these searches were a violation of the rights of the citizens. As a result of this intervention, the searches were discontinued. Perhaps if the attacks had been in the New York City subway system, the searches would have garnered more support.

The public will demand that we do all we can to provide the most protection with the least inconvenience. Training these aforementioned groups in the areas that were discussed can accomplish this mission. There will always be vulnerabilities in a free society, but the objective is to minimize these vulnerabilities by intelligence gathering and making people aware of the dangers. The limited amount of financial and time commitments required in adding these people to our national security force will be well worth the effort. The tools are available to us; it is now just a matter of utilizing the resources to our best advantage. Waiting for something to happen and then trying to take advantage of these resources is very much old-school thinking. In this case, we must work toward proactivity and not wonder why we failed to take advantage of an opportunity to make us more secure.

Endnotes

1. Time Magazine "Crimes of the century" www.time.com/time/2007/crimes23.html.
2. New York Times article entitled: "Virginia Tech shooting leaves 33 dead." www.nytimes.com/2007/04/us/16cnd-shooting.html. Published April 16, 2007.
3. "Timeline of worldwide school shootings," www.infoplease.com/ipa/A0777958.html.
4. Michael Elliott. "The shoe bombers world." *Time Magazine*, February 16, 2002. www.time.com/time/world/article/0,8599,203478,00.html.

Chapter 9

Politicians

Introduction

We have clearly demonstrated the need to cross-train responders in America in this book. Countless examples of areas of training, how they will benefit society and why, have been explored and examined. There is one other group of people who need very much to be involved in this process from start to finish. Their involvement can bring many additional assets, predominantly funding, to bear on the issue of cross-training our responders. This group of people is our elected officials, at the local, county, state and federal level. This group is generally overlooked when problems are identified and solutions are sought. But, if we look at things logically, it is the politicians at all levels who make critical decisions on funding. It is this body of civil servants who introduce, author, vote on and require implementation of legislative initiatives that affect each one of us. These are the very people who need to be there when we train, and when we identify problems. They need to see the hard work that goes into maintaining response annexes and to watch the efforts of our responders when they train. We need them to empathize with us when we are frustrated because we lack the required tools to get certain jobs done due to lack of funding. These politicians need to be aware of what we are able to accomplish and what we cannot do. I have already referenced the syndrome where people believe television to be representative of the real world. Politicians watch television too, and they need to be able to distinguish between fact and fiction. If we are afraid to invite them to our training and exercises, we cannot be surprised when our funding requests go unanswered.

Making Our Political Leaders Understand What We Face

Where do we begin trying to get politicians to fully understand not only the issues facing responders, but to act on these issues? Politicians control the abilities of responders to be in a position to be able to properly respond to any crisis. They have this great power as a result of their being the people who provide response agencies with their budgetary allowances. This process is followed at all levels of government, where budget requests are submitted and reviewed, and a budget amount is determined. That amount is sent back to the entity that made the request, and serves as the base of financial parameters for the agency or department to work within. Often, the budgets are voted upon in a vacuum, which is to say that the powers that be may not fully understand what is being requested, or why.

As noted previously, there is a trend in our society today to cut budget and staffing levels in government. While there are certain areas where this trend may be successful, the emergency response arena is not one of them. Our decision makers need to be fully aware of the ramifications their financial decisions will have. Fewer responders means slower response times, and sadly, this may result in loss of life or property. A simple solution to this dilemma is to involve the politicians in our training. Politicians all like to show up when there is a good story or a successful outcome to a serious issue, but they do not like to be at the podium when things have gone horribly wrong. Sometimes, the leaders of certain agencies are thrust into the laps of politicians, and in other agencies the politicians appoint the leaders to the key positions. Within a system, volunteer or career, response agencies and politicians must work together. The ramifications of hasty budgetary cuts need to be explained in painful detail to our leaders to ensure their ability to understand what may happen.

Helping Politicians Understand Our Funding Needs

To help the people who control the money to understand our challenges, we need to invite them into our world and allow them to see for themselves all that the responders are faced with. They should receive all of the training that first responders receive. I do not mean to imply that all politicians need to complete a police, fire and EMS academy, but the specialized training we receive after our basic training should be made available to them. If they were to take part in the training classes, they would understand the variety of threats we face and understand the time requirements to remain proficient in response strategies and tactics. So which areas of training will be recommended for these leaders?

Training for Politicians

CBRNE training is always an excellent place to begin training for anyone in the response community. These are the five most likely attack scenarios that will be

Figure 9.1 The Statue of Liberty is one of many structures that interest terrorist groups. Security at this type of structure has been greatly enhanced since 9/11. In mid-2009, the upper levels of the statue were made accessible to limited numbers of visitors for the first time since 9/11. An attack on such a national symbol would have tremendous psychological impact on our citizens.

faced by our first and second responders. Many people have no real understanding of what is involved in the CBRNE arena. Politicians need to be intimately familiar with each attack ideology, parameters, possible effects and casualty projections. To this end, we will review some of the particular attack methodologies and warning signs and symptoms associated with each type of attack.

Chemical attacks and incidents can be broken down into three distinct subcategories. These are irritants, nerve agents and vesicants. Irritants are the most innocuous of the three categories. These agents include mace, pepper spray, and the various riot control agents. Exposure to these agents generally produces no long-lasting medical effects. Exposure will result in tearing of the eyes coupled with a burning sensation that is designed to make affected people keep their eyes closed. A cough and reddening of the skin may follow, but again, these effects are not long lasting or permanent. The intent of these agents is to secure compliance from individuals who are resisting lawful orders or direction. History has shown these agents to be moderately to very successful in generating the desired effects.

Next are the vesicants or blister agents. Mustard Gas is the most recognizable agent in this subcategory, although an agent called lewisite is also in this group. Exposure to vesicants produces fluid-filled blisters on the body or in the lungs if inhaled. When the external blisters rupture, the body begins to dehydrate and suffer systemic failures of the kidneys and other vital organs. If the blisters are on the lining of the lungs when they rupture, the lungs fill with fluid, which can lead to death.

The final subcategory of chemical warfare agents is the nerve agents. The most common of these agents are sarin, soman, tabun and VX. All of these agents are very toxic and lethal. None is naturally found in the environment and they are therefore synthetic. In small doses, these agents produce symptoms very similar to those of the irritants. However, if a larger dose is experienced, the effects are far more devastating. They will include dizziness, difficulty breathing, seizures, respiratory arrest and death. Recognition of these signs and symptoms of exposure is critical if we expect to save lives. The differences between irritants and nerve agents are very important to both the victims, and the responders.

As for biological attacks, the bottom line here is that there are many different agents that can be used against us. These include toxins, viruses, bacteria and infectious diseases such as rickettsia. Response to these types of attacks will be delayed, as most biological agents require a period of time called incubation. What is of note with these agents would be the systems in place for early warning of an attack or outbreak of a biological nature. Learning how this system works and exploring the deadly nature of biological agents will be of great value to politicians who will realize the need for funding to prevent and respond to a biological attack. The manner in which biological attacks are monitored and investigated has already been discussed earlier in this book.

Radiological Training

One of the buzz terms in the post 9/11 era is *dirty bomb,* or RDD. A dirty bomb is a conventional explosive surrounded by radiological materials. Its general purpose is not to kill a great number of people, but rather to instill fear that radiation is present, and to disallow the use of a particular parcel of land until the radiation is removed. Politicians need to learn about all of the readily available sources of radiation, and why it is critical for us to provide adequate security for these sources. Let's delve a little more into the potential of a dirty bomb attack at a particular site, and the effects it would have on the people who reside in that area, as well as those in the state where the attack takes place.

Florida is a very desirable tourist destination for several reasons, including the climate, NASA and Disney World, among other factors. If a terrorist group wanted to wreak financial havoc in Florida without killing large numbers of people, a radiological attack would certainly meet those objectives. As with our malls, there are security forces on the job every day at Disney World. As guests enter, those with bags are subjected to a cursory inspection of their property, and those without bags are not subject to any search. If a group were acting in concert, each member could manage to bring components of a small explosive device—or several, for that matter—into the park. They could assemble the devices in any restroom or other semiprivate location, and have the devices detonate at a prescribed time. The devices could be left in garbage receptacles, in "parked" baby strollers or in briefly unattended bags. Once the devices were detonated, radiation would be released into the park area. The terror group could then alert the media, claim responsibility, and recommend that radiation levels be monitored at the park. If elevated levels of radiation were found, the park would need to be evacuated, tested for radiation levels, and all areas that are contaminated would need to be decontaminated. Naturally, word would spread through the media that the park had been subjected to a radiological attack. In the aftermath of this news, many tourists would be reluctant to visit, even with repeated assurances that the radiation has been removed and there was no danger in visiting the park.

How would this event affect the Orlando area, as well as the surrounding communities and possibly the state of Florida? If sufficient radiation was released, requiring the closure of the park—or parks if a scenario envisioned attacking multiple theme parks was employed—people would change their vacation plans. The Disney World employees would mostly be laid off for the duration of the cleanup efforts. Given the skeptical nature of Americans, the attendance at the theme parks would remain very low for a substantial period of time. This would have obvious detrimental effects on the parks as a whole. But what other industries are tied into Disney in Florida? Certainly hotels and motels would suffer due to lack of visitors. Rental car companies, airlines, tour bus companies, restaurants, gift and specialty shops would all see a dramatic loss of revenue. If the levels of radiation required the closing of the parks for as little as one month, the damage caused would be enormous. Vacationers would be very hesitant to return to the park area, for fear that radiation might still be present or that another attack might occur. I wonder if our political leaders have ever envisioned such a scenario? Had they been provided radiological awareness type training, this or a similar scenario may have been discussed during the training. This is just one example of one type of attack that would have serious and potentially long-lasting effects on our society.

Nuclear Attacks a Likely Scenario?

Nuclear attack scenarios are very unlikely to occur in America for various reasons. These include the redundant security features on most nuclear devices, the danger in trying to make a nuclear device, the relative unavailability of materials needed and a lack of knowledge in assembling a nuclear bomb. While there are many "suitcase," or "tactical nukes," missing from the former Soviet Union, the likelihood of a nuclear attack is very small.[1] Responding to a nuclear incident would be very dangerous for the responders, and extreme caution would be exercised to prevent more casualties. There are no protective ensembles that would protect the

Figure 9.2 Bridges and other critical elements of infrastructure are of interest to terrorists. Police are trained to be aware of suspicious activity on and near bridges and similar infrastructure such as gas lines, etc. Politicians have realized the importance of such critical infrastructure and have allocated tremendous funding to harden and secure them.

responders from exposure to the expected type and levels of radiation that would be present after a nuclear detonation. Our best defense against a nuclear detonation would therefore be to prevent it.

Explosive Incidents

The last CBRNE event to be discussed here is explosive incidents or attack scenarios. We have seen a great number of explosive assaults on our troops located overseas, in Afghanistan as well as in Iraq. In fact, many of our military casualties have resulted from IED such as roadside bombs. While this attack scenario has not been prevalent in America, there is always the looming threat, given the availability of explosive components and training manuals available on line that detail how to make and detonate a bomb. In addition to the IEDs, there are VBIEDs; there are suicide bombers and a host of other improvised explosive devices categorized by homeland security officials. The most devastating example of a VBIED on American soil was in April of 1995, when Timothy McVeigh packed 4,800 pounds of an explosive called ANFO (ammonium nitrate and fuel oil) into a rented truck and parked it in front of the Alfred P. Murrah Federal Building in Oklahoma City, Oklahoma. The resultant explosion killed 168 Americans and injured many others.[2] The magnitude of the explosion required the demolition of the building after the investigation was completed. This was the largest loss of life resulting from a domestic terrorist event in the history of the United States. It is interesting to note that most Americans fear an attack from Al Qaeda or some other international group, and yet the second-largest act of terrorism was perpetrated by an American citizen.

Suicide Bombers Are a Particular Issue

To date, America has been fortunate in that we have not seen any series of suicide bombings on our soil. Other nations of the world cannot make that same claim and have witnessed the devastating effects of suicide bombings. Suicide bombers are generally very successful in detonating when and where they desire, and accordingly, the casualties from these simple attacks are generally quite high. Suicide bombers have plied their deadly craft in subways and buses, as seen in London in 2005. They have also been very effective in tourist locations such as malls, social clubs, bars and nightclubs in many nations. While we as a nation are continuously exposed to the aftermaths of suicide and IED detonations, they have always been on another continent, and thus, we do not realize the peril with which we may be faced.

Our first responder communities should be provided with detailed training in the areas of IED and suicide bombings. Politicians need to be exposed to these training classes as well. They need to see the effects of the bombings on victims, as well as the community. They need to be fully aware of the threat that these types of attacks pose to Americans. The training provided to security forces in other countries needs to be shown to our political leaders so they can see exactly what training should be mandated and provided. It is one thing to be exposed to a short blurb about a suicide bombing while watching the evening news over dinner. It is however, entirely different when you are sitting in the class, are exposed to crime scene photos, and are able to witness the devastation. In this case, a picture is worth a thousand words, maybe even more. Seeing is believing, and political leaders need to see exactly what is involved after a suicide bombing or an IED detonation.

Terrorism Awareness and Indicators

While CBRNE training is an excellent place to start, there are still other areas where politicians need to be trained. As with all of the other responders covered in this book, terrorism awareness should be provided for politicians. This broad class presents us with a history of terrorism and a view from the side of the radicals. When studying the history of the Middle East region, it is easy to see how clever terrorist leaders are able to manipulate devout religious people into believing a set of facts

Figure 9.3 The Supreme Court Building in Washington, DC, is another national building that could be of interest to terrorists. The U.S. government has always employed a great deal of security at structures such as these, which are open to the public. As with the Statue of Liberty, a psychological toll would be a goal of the terrorists along with the potential for great loss of life.

that have no basis in reality. Awareness training also explores how terrorist groups are funded and gives examples of techniques that can be tracked and prosecuted.

One such technique involves the fraudulent redemption of coupons. In this type of scam, merchants of tiny stores and bodegas present high volumes of coupons to manufacturers for redemption. In conducting investigations, it is apparent that a small store is unable to have sold the number of goods they allege when submitting coupons for redemption. The coupons have no real cash value for consumers, but, when submitted to manufacturers, they become very valuable. Coupons clipped from magazines and newspapers that are not redeemed, but are forwarded to manufacturers, generate large cash flows for unscrupulous vendors. The revenues garnered from this illegal and fraudulent coupon redemption scam are funneled to terrorist groups overseas. Unwittingly, the manufacturers of goods in America are providing funding for certain terrorist groups.

Other areas of terrorism awareness include attack methodologies and some exposure into surveillance and counter-surveillance tactics and techniques used by terrorist groups. Exposing politicians to this type of training will enable them to see the inner workings of our terrorist enemies, and they can see the value of expenditures in the intelligence communities to help infiltrate these groups and expose them for what they are. Funding intelligence and enforcement personnel can reap great rewards if we are able to prevent such crimes from occurring. The elimination of these types of funding sources for terror groups will be of tremendous value to our law enforcement communities when trying to protect our citizens.

Politicians at Our Drills

Other training options for politicians come mostly in the form of being observers at drills, tabletop scenarios and exercises of annexes of Emergency Operating Plans (EOPs). By inviting the political leaders to these events, they are able to see for themselves how tedious, taxing and demanding the responder training can be. They will begin to empathize and sympathize with those who serve and protect. This exposure may enhance the ability of responders to receive increased financial support from their elected leaders, which will make their jobs much easier. Politicians are no different from the general public in a sense. The general perception about exactly what a typical day is for our responders is far away from the reality. Providing a first-hand look into the challenges faced by responders will benefit the politicians, who in turn can explain the real world of response life to their colleagues. In the remainder of this chapter, several of these training scenarios will be explored, along with issues related to funding.

Biological Attack Scenarios and Training

In addition to CBRNE training regarding biological agents, politicians should be expected to attend any drills that test the jurisdictional response to a biological attack, or any testing of the annexes for such an event. For them to witness the complexity of responding to a biological incident will clearly open their eyes to the pressures and demands faced by first responders. Having these leaders of our communities sit in on the implementation of these plans and to review the medication distribution plans will have great impact upon them. A practical evaluation of the response annex will take tremendous planning, coordination, communication and cooperation among numbers of agencies. Local testing will impact the county plan, since the local entities will receive their allotted medications from the county. This in turn will require activation of the state agencies, which will make the necessary arrangements for the counties to receive their medication. Obviously, the federal agencies that control access to the medications will also be affected, and will be required to participate in the drills or exercises. Coordination among all of these agencies will be crucial in times of actual events, and these drills allow them an opportunity to meet and work with each other.

Reverting back to exercising this annex on a local level, there will be tremendous amounts of paperwork, and also there will be logistical issues relating to the

POD and how the medications will be retrieved and distributed. Security concerns for the stockpile of medications will be an issue from the time they are received by the county until they are all distributed, or returned to the county. Additionally, there must be a system in place that will allow persons other than medical doctors or pharmacists to dispense the medications. What might seem like a simple testing of an annex may turn into a prolonged and difficult-to-manage event. Politicians will see how incident managers function at this type of drill, as well as gain a real sense of what manpower and resources would be needed.

If the politicians are from the county government, they will have an even better sense of what is involved, since there will be many county assets brought into play. The state level politicians will be amazed at the complexity of the operation and the amount of resources required to handle such an attack response scenario. Having witnessed the testing of a similar annex, politicians will have a better understanding of funding requests for certain materials and equipment by their responders. In this scenario, there will be many different players involved in the drill from different agencies and jurisdictions. They will include, but may not be limited to, the following: public health, OEM, law enforcement, incident managers, volunteers and similar representatives from county and state agencies. All of these agencies must work together and strict adherence to the parameters in the tested annex must be complied with.

Observing Chemical Attack Training

Another area where political leaders can benefit as observers would be during training related to a chemical attack. This type of training, as with biological response, will certainly raise the awareness of what is required at an incident involving chemical attacks. Several years ago, my county conducted a large-scale drill that simulated a chemical attack on an enclosed mall of relatively small dimensions. The planning process for this drill went on over a period of many months and involved planners from local, county, state and federal agencies. During the drill, the owners of the mall allowed access for all of the participating agencies. The drill was conducted very early in the morning, so as to not interfere with the operations at the mall. In an effort to speed things up, much of the equipment required was set up in advance. Staging areas were pre-determined, a mobile command post was brought to the mall, and safety, accountability, decontamination, triage and other areas were operating the minute the drill was started. There were local girl scouts and their parents, who volunteered to be used as victims.

Once the drill began, federal agencies, along with state and county responders, served as moderators and observers for the drill. The simulation was that of a chemical attack in the thirteen-screen cinema, which has a mall entrance as well as a private entrance from the exterior. My fire department was participating, and reported to staging in accordance with the plan. Eventually, we were required to report to the incident location for further assignment. Once we were checked in, my crew and I were assigned to transport the victims from the decontamination line to the triage area. There, the patients were further assessed and designated for transport, based on the severity of their conditions in accordance with governing triage protocols.

Shortly after the victims were removed, my crew was asked to enter the building to conduct a search for a missing security guard, presumed to be in the theater area. I inquired as to the identification of the chemical agent and was told that there was no positive identification. I therefore proceeded to refuse to allow my crew to enter the theater area for fear they would not have adequate protection. This refusal led to a short "discussion" with drill organizers. I stated my position that a lack of product identification created an unsafe environment for my crew, specifically as it related to PPE worn by my firefighters. I was then advised that the drill could not end until this last victim was accounted for. I requested that drill organizers note that in a real-world scenario, my crew would remain outside. With this objection clearly noted by drill moderators, and in an effort to end the drill, my crew and I entered the theater to locate the missing security guard.

Immediately upon entry, moderators advised the crew designated to search the opposite side of the theater that they were now victims of the agent. My partner and I were not stopped, and we located and removed the last civilian victim. As we were preparing to leave the drill, I was stopped by the drill moderators and observers and asked to provide input. My initial report was that there were not enough medical resources at the scene, but this was expected. My initial refusal to enter the theater was mentioned, and a local politician inquired why I had refused to enter. After hearing my explanation, the politician indicated that he was very surprised to learn that firefighting ensembles offer virtually no protection against chemicals.

This real-world scenario reinforces my opinion that allowing—or insisting—that politicians participate or at least observe first responders in action allows them to learn what it is that we do and the challenges we face. It was interesting to note that several other firefighters present during my debriefing were unaware that their fire gear was ineffective against chemical agents. Certainly, in this case, having political leaders present opened some eyes and alerted them to areas of chemical attacks of which they claimed to have no knowledge. The primary area of questions from politicians revolved around the need for decontamination, as well as decontamination procedures. This fact, too, was valuable in having them understand what equipment is requested and why it is so badly needed. Certainly, there is a benefit to be gleaned from allowing and having our elected leaders participate at our drills and training sessions.

Attendance at Table-Top Scenarios

There are still other training scenarios where the presence of politicians can be beneficial to their understanding of our operations. In instances when planners are attempting to save money and manpower, tabletop scenarios are an excellent option to meet those objectives. In a later chapter, we will provide information on how to set up and carry out a tabletop drill scenario. Briefly, this type of training or form of exercising an annex is designed to limit manpower and resource requirements and reduce costs associated with the completion of exercise. During this type of training, a scenario is chosen and roles are assigned to all participants. In exchange for using personnel, a technique referred to as *modeling,* or *simulating* is used. This concept allows for participants to claim that in a real-world event, a certain number of responders would be assigned to complete certain tasks. During the execution of the tabletop drill, those responders are not actually present. Only the planners need to be present during the actual exercise. These types of training still require tremendous amounts of advance planning. It is also desirable to invite representatives from other agencies to observe and provide independent feedback after the exercise is complete. Having political leaders present during the actual drill will afford them an opportunity to observe the training, without having to foot a large bill for resources and manpower. Further, the politicians can observe the problem-solving abilities of their first responder leadership in a safe, quiet environment. Allowing them to witness the drill may aid them to understand potential issues that may be encountered during an actual emergency. This is one benefit of their presence during exercises.

Trade Shows and Conferences to Reinforce Training

At this point, we have trained the politicians and have allowed and encouraged them to attend our drills, exercises and tabletop training sessions, but can we do more? We most certainly can. We need to encourage the politicians to attend conferences and trade shows. Conferences will generally have speakers who will address certain *hot topics* in their respective fields, or they will provide training in the particular response area. Additionally, at many of the conferences, trade shows are going on simultaneously. These trade shows will generally have new products on display, which provides you an excellent opportunity to discuss new and evolving technology with your bosses. Also, existing technology is available for purchase at reduced pricing, which may bode well for you and your agency if a strong enough case is made for the purchase. Having manufacturers' representatives on hand to demonstrate the desired product and answer questions will be of great benefit. These sales representatives will be able to explain in great detail how the technology or product can be of value to your jurisdiction. Having an independent person speak to your leaders can serve to reinforce your position as to why a certain piece of equipment is needed. Lastly, conferences enable people to interface with other responders, and possibly those with issues similar to those they are facing. Allowing your political leaders to see that your problem is not unique, and that others in the same field have similar issues, may make your position for funding stronger. Additionally, the opportunity for your leaders to meet with other leaders, both political and within the response community, can pay

Figure 9.4 Ferry boats like this one are also of interest to terrorists for many reasons. These boats are capable of carrying vehicles and passengers. Security is not as stringent as in commercial aircraft, and they carry large numbers of passengers at peak times each day. Government has sought ways to improve security on transportation systems such as these.

huge dividends. The potential upside of bringing them to these conferences is very large, and they should learn a great deal about your issues.

Having invited political leaders to be trained, and then to observe our training in action, what is the benefit we hoped to achieve? The first benefit should have been met. By interacting with us these leaders should have a much better understanding of what is required of us in cases of emergency, the time and effort we must expend to prepare ourselves, and the issues and challenges we face as responders. But, realistically, the hidden agenda was to allow the people who control the money to see what we are up against. Then they can envision a particular piece of equipment that we may need and understand how it is of importance to us. The real reason we invited them into our world was to encourage increased funding of our agencies, including increased allotments for manpower. All emergency response agencies know there is intense competition among all response agencies for precious budget dollars. By exposing the people responsible for the division of the financial pie to your specific environment, you would hope that the wave of goodwill would be extended to your budget. If our leaders cannot empathize with us, there is virtually little hope of an expansion of your budget. More importantly, in tough economic times, dollars for responders are getting more and more precious. As indicated earlier, since we have been fortunate enough to not have been victimized by a terrorist attack, the perception is that we have done enough. The people who determine your budget may also share this perception. This would be a dangerous position for politicians to endorse. We must retain the current level of operational readiness, and in fact, we must forge ahead. To accomplish this mission, we need adequate funding and support from our leaders. For them to truly be on board with your operational vision, they must be able to see your problems through your eyes.

Federal-Level Politicians

On the federal level, there has been an infusion of money into the coffers designated for emergency response—and homeland security as a whole. This infusion of funding has had a dramatic effect on the fire service in the United States. For years, there was little grant funding available for the fire service, and this forced many fire agencies to do without tools, gear and equipment that were desperately needed. The tragic and sudden loss of 343 firefighters on September 11, 2001, in New York City focused the spotlight on the fire service and the inherent dangers faced by these brave responders on a daily basis. Congress was quick to address the deficiencies within the fire service, and enacted the Fire Act. This federal legislation created a competitive grant program for the fire service that was initially funded at $750 million.

The grant had a matching component, and these matching funds were tied directly into the population each entity served. Currently, the lowest match is at 5 percent, which is reserved for the jurisdictions serving the lowest populations. The grant outlined certain areas where agencies would be able to apply for funding. These areas included firefighter safety, operations, vehicle acquisition, etc. Many departments applied for these federal dollars, and the gear, equipment and tools were purchased by agencies and placed into immediate service. This program allowed firefighting entities with little money to replace aging, noncompliant equipment with newer state-of-the-art equipment. The program has faced some serious financial cutbacks, but it is still expected to be funded $350 million in 2010. Additionally, this program was expanded to include emergency medical service responders as well as the fire community. This was a good idea, as the medical responders have significant funding issues as well.

But what does the future hold for the funding of our various first responder entities? Can we expect the funding levels seen now to continue, or will we see a decline consistent with the overall economy? Financially, America is in trouble now, and the budget axe is liable to fall anywhere support dwindles. Many fire and first aid squads are still in need of resources, and they are not in the position to purchase them alone. Given these tough financial times, certain grants are being cut or eliminated completely. Grant opportunities for the hiring of law enforcement and fire personnel have been slashed or completely shut down. But, is it solely the economy, or are there other factors at work?

Maintaining Funding for Responders

A dangerous scenario for the American response community would be a loss of current funding levels for any reason. Budgets are always open for debate among the political powers that be and fluctuate annually. Any disruption in the flow of money for first responders may have dire effects on their ability to provide service at the levels to which their clients have grown accustomed. Tough times require tough decisions to be made, and sometimes the choices that are ultimately made are not the right options. Complacency has a tendency to set in, and in certain jobs, this factor can prove deadly. For firefighters, it is the same false alarm at the same building, until one day it is a fully involved fire and the crew is not ready to battle the blaze. For law enforcement, it can be the *routine* traffic stop that turns deadly with a wanted felon at the wheel. For politicians, complacency can rear its dangerous head when constituents determine that since we have had no terrorist attacks in America since 2001 that the threat has gone away. Believe me when I tell you that the threat has not gone away, and in fact, is as grave as ever.

Slashing budgets in response to rhetoric surrounding a lack of successful terrorist attacks would be a huge lapse in judgment. No one can be certain when, where or in which form the next terrorist attack will occur. We have put forth a mighty effort to become better prepared for the next incident, and we cannot afford to lose focus on the issue this late in the game. Funding was provided, and that funding provided responders the ability to lay the groundwork for preparedness. To turn away from the good work started years ago would be tragic should our responders lapse back into a pre-9/11 mentality. Much of what was purchased needs to be maintained, and skills need to be practiced or will be lost. Funding needs to be maintained or increased to allow for the progress made to continue. We cannot allow the skills to regress or have the equipment fail to be maintained.

Americans need to keep the pressure on the terrorist groups overseas, and we need to continue to fund our first responders at home. We cannot afford for any of our leaders to waver in their support of our responders, and in fact, we need to enhance, and renew our commitment to our first responders. We need to pay close attention to how our elected officials choose to spend our tax dollars. Cutting funding for protecting our families in favor of *pet projects* or similar proposals cannot be tolerated. Americans, especially those in the first responder community, need an accurate accounting of the voting patterns and habits of our elected officials. We cannot tolerate lies, misunderstandings or plain old deceit. The tools that responders rely on to get the job done are very costly, and we cannot afford cuts to our budgets. The American first responder is without peer in the civilized world. The bravery, courage and dedication shown by these responders on a daily basis is testament to this fact. I refer to the more than 400 responders counted among the dead in New York City on September 11, 2001. To suddenly reduce funding would be an insult to the memory of those heroes, as well as the memories of every other victim of that horrible day. We need to take our leaders to task when they are deciding where to spend our money. We cannot allow special interests to take away our resources. First responders and their families need to monitor the actions of political leaders when it comes to funding.

Summary

The importance of politicians in the response community has been clearly demonstrated. These are the people who are entrusted with the responsibility of allocating our tax dollars. They need to be trained in the same areas as our first responders, and be exposed to what the responders are exposed to. They should have training in all of the CBRNE attack scenarios. In this manner, they can have a basic understanding of what challenges will face first responders in the event of another attack. Knowledge is power, and providing the politicians with this knowledge will empower not only them, but us as responders as well. Their experiences in the classroom environment will prepare them for the next step in their ongoing education.

After we have provided the classroom training for the politicians, we need to follow this up with hands-on experience for them. Requesting their presence during any training you are conducting can certainly leave an impression in the mind of the politician. Large drills examining a variety of responses will provide an in-depth look at the complexities involved in incident management issues at a terrorist event. Observing how all responders must work together and work with county, state and federal partners will also provide a level of understanding to which many political leaders simply cannot relate. Complex situations require complex management strategies, and politicians need to realize this. In observing the drills, they will not only learn what our responders are capable of, but, more importantly, what they are not capable of.

Real-world incidents are nowhere near what they are portrayed as on television and in the movies. That is why television shows and movies are called entertainment. In reality, police officers are not involved in running gun battles several times per week, no matter where they work. Additionally, if they are involved in a shooting, they are certainly not out on the street later that day. Similarly, structure fires present firefighters with zero visibility conditions, high heat environments and the possibility of disorientation or injury from structural collapse. If firefighters were in a structure fire without breathing apparatus, they would die, just as other victims do. EMS workers do not save every victim on whom CPR is applied. While it is admirable that entertainers try to portray responders as a group of heroes, they may have unwittingly hurt us in the long run. Public expectations have become substantially over inflated. We cannot accomplish in the real world what some of these responders are able to accomplish on television and in the movies.

Allowing politicians access to our training exercises can help to dispel many of the common misconceptions regarding our abilities. This too is extremely important, so that our political leaders have a more grounded and realistic idea of our capabilities. Politicians need to be aware of the costs of our gear and equipment, and to understand the reasons for the cost. Requesting that they attend training conferences and trade shows will certainly allow them to see the high cost of our equipment. While at these conferences, they can make efforts to meet responders and evaluate the issues facing them in relation to the issues facing responders in their own communities. Hopefully, there will be other politicians present as well, and they can discuss the many issues that each community faces. They will also have a realistic concept of the cost of our tools and will not be so shocked when they are presented with proposals for new materials. Exposure to responders from outside of their response area may tend to confirm pleas requesting tools, training and equipment. This also may be of immense value to each agency.

With all of this in mind, we would hope that the politicians in charge of budget allocations will be on our side. This may or may not be the case. If they choose not to support our needs, then we need to take them to task. They will need to be reminded of the consequences of becoming a reactionary force, instead of a proactive force. We cannot allow complacency issues to invade their thought processes. Groups such as Al Qaeda will not plan their attacks based on the calendar. They are meticulous in their planning, and expend great efforts to make their training as real as possible. They train and train and train until all of the members of the team are working like a well-oiled machine. Let us not forget that it was over eight years between the 1993 bombing at the World Trade Center and the 2001 airplane attacks. We must recognize the patience exhibited by terrorist groups and be ready for a long battle.

The only thing close to a trend may be that terrorists have carried out the last large-scale events within months of the country's electing and seating a new president. President Clinton was newly in office in early 1993, and President Bush was in office less than a year in 2001. Therefore, if complacency is allowed to enter our elected officials' thought processes, we need to remind them about history. If the politicians are not receptive to the needs of the first responders, we need to call them out. If that does not work, we may need to explore different options for their political seat the next time they are up for reelection. Politicians can be a tremendous asset for us, but they can also be an equally dangerous liability. We need to know where they stand on our safety and first responder issues. If we are not satisfied, we can always seek an alternate solution. We must always remember that the government is made up of people, by the people and for the people. If our elected leaders forget that, we can certainly find others who will be willing to embrace that concept. We need our leaders to do what is right for all of us, and to ensure that we are as prepared as possible to face our next challenge. We have the power to make sure this is what actually happens. If we allow it to go undone, we are the ones to blame.

Endnotes

1. Office for Domestic preparedness course entitled "Weapons of mass destruction (WMD) radiological/nuclear awareness", Student Manual pages 4–7.
2. "Emergency response to terrorism: Basic concepts" course, offered by the National Fire Academy. Student manual pages 1–3.

Chapter 10

Training Opportunities, Grant Tips, and Related Issues

Choosing an Instructor

Now that we have identified our target audiences for the concept of cross-training, we need a plan to make the information readily available to all who want the training and all who will receive it. Where do we begin in the search for competent, qualified and knowledgeable instructors to provide our identified responders with this newly desired material? Several sources are suitable to provide training for each responder who is in need of further education. The chosen educators must be qualified to instruct and must bear some form of certification, or accreditation, from a recognized body.

Choosing the right trainers is just as important as choosing the correct areas in which to be cross-trained. Trainers must have a basic understanding of what the trainee does on a daily basis, and be in a position to relate how the new training can be best utilized and applied by the new student. There are great numbers of excellent instructors who have years of experience and the proper background who are able to provide this training to the targeted audiences. These individuals may work as instructors at police training facilities and fire academies, and as EMT educators etc. Further, there may also be qualified instructors in police, fire, emergency medical response and hazardous materials departments.

Often, police and firefighters, as well as hazmat technicians and medics and EMTs, will also serve in a volunteer capacity in another field. These people, who already have a broad base of cross-training and experience, can be an excellent choice to conduct this training. For example, when my department looked into certain types of training, the instructor chosen to present the materials became an issue. It was decided that since I had the ability to present the material and was able to relate exactly how our responders would use it, I would be the instructor. An outside instructor may not be familiar with our policies and procedures, daily duties and abilities of certain employees, and accordingly, the training may be lacking in those areas. The importance of this "inside information" can be of paramount importance. Later in this chapter other instructor and class or course options will be presented. These resources are trainers and classes that can be obtained through the federal government, and who will bring the training and materials directly to the agency that has requested the training to be presented.

Instructor Basics

Many times, the ability to have instructors from your agency available to conduct the training will make the training process go much more smoothly. This is especially true if the chosen instructor is well received or well respected by the target audience. But what if the instructors from your agency are not able to present certain training to your other employees? If you have certified or credentialed instructors on staff already, the solution may be quite simple.

For most of the training recommended, there are train-the-trainer (TTT) course offerings. What these offerings consist of is an experienced instructor in that particular area who will attend a class designed for other instructors to be certified to instruct in the desired area. The TTT offering will allow the veteran course instructor to review the course of instruction with other general instructors who will then bring the class back to their agencies to be presented. The veteran instructor will offer training tips for the new instructor at the TTT class. These pointers will cover the entire class to be offered, and will include tips on scenarios, presentation methods, testing and application of the new material etc. These TTT classes give a new instructor significant guidance in how the materials may be presented to ensure that all of the relevant aspects of the class are covered. The experienced instructor can relate to the good and bad points in the course materials, and can point out pitfalls inherent in the materials and presentations. Additionally, they will provide examples of how scenarios can be conducted and how to best approach group activities if they are a part of the class offered. Once the new instructor has taken the TTT offering, he or she is then ready to prepare him or herself to present the course to new students. It is often desirous after a TTT is completed for the new instructor to sit in on an experienced instructor as the material is presented. This will allow new instructors an opportunity to see how another instructor presents the information and will allow them to observe time and course management.

Additionally, this process can serve as a refresher training session for the new instructor. Prior to the new instructor's presenting materials to a class for the first time, the entire course and its presentation method should be reviewed. This process will allow new instructors to find areas in the program where they may be lacking in knowledge themselves, and they can then make inquiry of veteran instructors or subject matter experts for guidance. Once new instructors feel quite comfortable with the materials to be presented, the process of training others from their agency can begin in earnest. Once the materials are presented, the first few classes of students should be given forms to evaluate the course materials, and more importantly, the new instructor. These forms should be filled out anonymously and be fairly in-depth. Experienced instructors can review these performance evaluation forms, and make recommendations for the new instructor. The instructors should review any issues regarding the performance of the new instructors, and strategies for correcting these issues can be developed and implemented. After the adjustments are made, the class should be getting better marks from the students. Once a new instructor works out the issues documented by the students, the next batch of students will appreciate the instructors' efforts. This then makes that instructor one with whom students will enjoy working and someone they will trust to value their input. An instructor with a good reputation for presenting interesting classes in an interesting manner will get the respect of the students. When this type of instructor makes another offering, participation and retention on the part of the students will be much better, and the full value of training will be achieved.

Outside Training Options

If your agency does not wish to have materials presented in house, what are your options for training your members? Often, having training located at an off-site or neutral location can be beneficial. There will always be veteran staff members who do not feel that they need new training. Placing your staff in an unfamiliar place will make them reluctant to act out, since they are not on their own familiar turf. The obvious answer will be to utilize facilities that are designated as training academies, or locations where continuing education is provided. Police, fire and EMS training academies are not dedicated solely to students who are recruits, or rookies, and therefore are an excellent option as a training location.

Each responder discipline requires some types of refresher, or in-service training. Dedicated training academies and facilities are an excellent resource for cross-training venues. Each type of facility will generally have *subject matter experts* on staff who can present training classes for your responders. For example, if you desire to train police officers on what actions are acceptable, safe and practical for them to perform at a fire, you can send them to a dedicated firefighting training facility. While there, fire service instructors can lead them in a course of training designed to address these very issues. In fact, if there is not a program readily available, many training facilities can develop a specific program to address specific training needs.

For example, as already noted, leaving doors open during fires may lead to more intense fires, may cause fire spread and reduce the likelihood of successful victim rescue by fire crews.

A fire service academy can be approached by law enforcement, and asked to develop a training course that addresses all of the issues police officers will face at structure fires. The material can also present one or more safe strategies and tactics that can be attempted by police personnel. Once the training facility develops the program, law enforcement will be trained and have a far greater understanding of the science of fire, and how best to safely and effectively respond to fire incidents. Also, if the training program is one that is well developed and received, a petition may be made to have that instruction added to the basic training of other responders. The same scenarios would be present for fire and other responders to be trained in other disciplines, such as terrorism awareness, attack methodologies and tactics, suicide bomber issues etc. Using existing training systems and facilities is an excellent strategy for those who do not staff instructors with the ability to cross-train their own responders in multiple response disciplines.

There are still other training options for those who need to be trained in multiple responder disciplines. If you have no instructors on staff, and would prefer not to report to dedicated training facilities, you can contract with civilian trainers. In certain cases, there are distinct advantages to contracting civilian trainers. They will see your personnel as clients, and as such, they will be judged on their own merits. There will be no potential conflicts in having responders from another response discipline coming in to train your staff. Most often, these private contractors are very well trained, and have real-world experience in the field, which they are contracted to instruct. Their training, experience and education are usually enough to get the approval of the students, who will be more accepting of a person whose area of expertise is being presented to them. The civilian instructors can be contracted and brought to your facilities at times that are most convenient to you and your employees or volunteers. Some of the instructors will charge a per student fee, while others will charge based on the materials presented and the time required for the class. However the classes are billed, there is greater control over the instructor and the delivery options.

Civilian instructors are much less likely to have prior knowledge and contact with your agency, and sometimes a detached instructor with a certain area of expertise will be better received by those who are in need of that type of specialized training. For example, if you want your responders to become familiarized with certain chemicals produced and stored in a facility in your area, civilian instructors may be the best option. An employee of the facility should be able to review existing response protocols, specific characteristics of the chemicals on their property and other dynamics of the facility or products contained therein. This person should be very well versed in all areas of the operation of the facility, and this level of experience and knowledge will benefit responders. Should you want to educate responders on subtleties and nuances of the Muslim religion, customs and traditions, an Imam or other religious leader would be an excellent option as an instructor. Whichever civilian instructor is chosen is of no matter, so long as the person is qualified and willing to provide the training. Some responders will relate better to another responder providing training, while others will be more attentive to a civilian instructor. Each agency needs to determine the best delivery option.

Other Training Resources

Other training options remain for most agencies to consider if the aforementioned are not ideal for their agency. Again, the best instructor is the one who has experience, training and a desire to provide the training for the agency. As previously indicated, our military personnel are already cross-trained in a number of areas. The military has done a fantastic job of preparing its troops to be deployed into various hostile regions and to respond to a variety of environments and political climates. The military does its own research in a number of areas and has very qualified, competent and capable trainers. If there are military responders and trainers in your area, they may be willing to provide some cross-training for your responders. Since their branch of service is already paying them, there should be no cost for the trainer to report to your agency and train your personnel. The military personnel are very well trained in areas such as IED and certain CBRNE areas such as biological and chemical weapons. This type of experience and training can make the military responders an excellent training tool. As stated before, instructors with real-world experience will generally have the respect of the student, and closer attention will be paid to what these responders offer. Given the publicity surrounding the use of roadside bombs, the news of suicide attacks in the Middle East and possible exposure to chemical and biological weapons, military personnel are very attractive options as instructors. So long as instructors are willing and able, they are a valuable tool to be used in attempting to cross-train civilian and other professional responders in certain areas.

Making Responders Want Training

Having the ability to locate and bring trainers into any location is a big step in beginning the cross-training process. The other obstacle faced by administrators regarding training in new areas is having a willing audience. The people and responder groups who have been identified as being in need of cross-training bear tremendous responsibility as well. The group of students must have a desire to be students in the training class, and act accordingly. As an experienced instructor, I can attest to the fact that students who are in a training class only because they *are required* to be there are generally not good students. Their attention span is very limited, and they may have a tendency to distract those around them with under-their-breath comments and negative body language. These students need to be dealt with early on, prior to their contaminating the learning environment for those students with a genuine desire to learn. Therefore, the students in each training class must truly either want to be in the class, or must at least understand why it is that they are in the training. Reluctance to accept new training ideas and concepts can hinder the learning process almost to the point of futility.

Cross-training the reluctant student requires a competent and crafty instructor. The instructors must find a manner with which to make a connection with the student. In my experience, presenting the worst-case scenario for failure to recognize the concepts of the new training area is a certain way to get the interest, if not the attention of the students. For example, when instructing on hazardous materials and the possible lethal consequences of lack of training, it is very easy to get the attention of tentative and reluctant students. Phrases such as "at least you get your first helicopter ride," or "if you screw this up, you just end up dead" have always managed to get the attention of the students. Also, pointing out how and where use of this training might be encountered when they are working can also serve to increase their focus on the topic. Simply forcing training on people who have no urge, interest or desire will generally not produce the desired effects. It needs to be stressed that the goal is not to certify anyone in another competency, but rather, to increase awareness in certain areas for the protection of the responders as well as the public they serve. If we are able to increase the interest of the potential students, training will have a far greater and wider-reaching impact than if we were simply to force it on the responders.

Another technique the experienced instructor can utilize to help students become interested in the new training area would be to reinforce their importance regarding the new training. Often, students or veteran responders will not see the need for training in new response arenas, and will need to be coaxed into buying into the necessity of the training. Adeptly explaining why the training is being given, and detailing how it will benefit them may draw these veterans into seeing the potential benefits of the training. Human nature is very fickle, and when people are in positions where they are not appreciated, or if they feel they do not matter, their ability to focus suffers. When the importance of individuals is stressed and reinforced, they become much more attentive, and in my experience, will become better note-takers and will often ask more questions. Providing training is not enough; the materials need to be absorbed, processed and retained. Those who are getting the training must see a benefit for the time they must invest in taking the prescribed training.

State Police as Trainers

There is yet another source of training that is sometimes overlooked completely by most agencies and administrators. Most of the states in America have well-trained and experienced state police officers, troopers or highway patrol officers who would be able to provide training for other agencies. Since the jurisdiction of these state agencies is quite large, the state agencies will have many different units or sections designed to address certain issues. Many of these issues will result in additional training for the state officer, and if these officers are trainers themselves, they can then bring the training to other responders. State law enforcement agencies will most certainly be willing to provide their personnel for training others within their state who will need specialized or cross-training.

In Chapter 3 of this book, we asked, "why can't we all just get along?" Sometimes there is friction among law enforcement agencies on different levels. If this is the reason that agencies are reluctant to utilize state agencies for training, then we must focus on the need for training and get past our petty differences. There are resources that can be put to good use and trivial differences need to be overcome to serve the greater good. State agencies have larger staff and more resources and can provide insightful training to any agency that is in need. Failure to include these trainers in the training scheme is not an acceptable situation, especially when the training is much needed and there are available and willing trainers with excellent qualifications and experience.

In-House Instructors

The final resource that agencies can use when prescribing or implementing cross-training is those in-house who are already cross-trained. Many professional responders also serve in a volunteer capacity, either within the same discipline or in another response discipline. These cross-trained responders can be a vital tool in the ability of an agency to get members to want cross-training. There will certainly be good-natured ribbing between people who work in one area of response, and volunteer in another. However, this same bantering can produce a desire on the part of those in need of training to realize the need and to endorse the program. Whether those already cross-trained are the instructors is of little consequence in the final analysis. The ability of these individuals to suggest why the training is needed may serve as the impetus to get the interest level rising in potential students.

Many responders who work across multiple responder boundaries are well respected by their peers. In fact, in many scenarios, those cross-trained will be sought out, requested to be dispatched or sent to certain incidents. These responders can discuss the practical real-world applications where cross-training can be very beneficial to members of their agency. Once the thought is in the head of a potential student, half the battle is won. Having peers who can relate to the challenges faced by other response disciplines and offer solutions for rarely encountered problems will allow for an easier transition in training from one discipline to another. If there is reluctance on the part of those in need of training, others with the training already need to step up and explain why the training is needed. This must be done so that reluctant students can visualize and accept the proposed training. A member who can list specific incidences where the cross-training was beneficial will garner support from his peers in seeking to justify and receive the training. People are much more apt to accept something when they can see intrinsic value to them on a daily basis, or where it can be beneficial at least periodically. It is imperative for employers and administrators to use the resources in their own agencies to their best advantage. Allowing these resources to go untapped is foolish.

Mutual Respect

Trainers and trainees must have similar expectations of each other and certain similar expected outcomes for the training. If both the trainer and trainee are not on the same page, the full benefits of training will not be realized. A genuine desire to learn must be instilled, and the training must be conducted in such a manner so as to allow students to visualize how they can implement the training during the routine day-to-day carrying out of their prescribed duties. The trainer's hope is that the students will be interested in the subject matter and that they are hoping to get something out of the training. Students expect that the instructor will be prepared, on time, experienced and well versed on the subject matter being taught. If we make expectations for the students, we must realize that they will also demand certain things of the instructors. There is nothing worse than taking a training class with an instructor who is unprepared or has no practical understanding of the materials being presented. Both the instructor and student need to be well prepared for their role in this critical exchange of information.

Within the response community, a number of agencies, entities and vendors are trained in various areas of responder training and will provide discipline-specific training. Some of these are from accredited training institutions, some from local and state training agencies and others from federally funded programs. Regardless of which area of expertise is needed, there is probably an entity that can provide the training needed. In some cases, the vendor or agency will require that the students come to their location, while the majority of others will come to your location to provide the requested training. Some of the training locations that require the student to report to their location can have the costs of the student paid for by the federal government. The cost of airfare, rental cars (if needed), food and lodging are all subject to payment by the federal authorities' regulating and approving the training. These training programs allow the student to relocate to a specific area where certain training topics are given. For example, in Nevada, approximately one hour outside Las Vegas is a training area dedicated to radiological and nuclear training. Students are brought to the Nevada Test Site and remain on the designated campus for several days. During their stay, they are given classrooms and practical training exercises. They are trained in the use of certain radiological-detection equipment, decontamination and entry procedures. The students stay in dormitory-type housing and are fed three meals per day. For classes such as these, there is prerequisite training generally that must be documented for all students. Further, attendance is strictly enforced to assure that students get the full benefit of the training experience. These on-site classes will routinely require a test be taken prior to credit being given for the training. The benefit of this type of training is that the student is brought into the field and can apply what is learned in a live, yet controlled environment. This type of practical hands-on training is very beneficial in reinforcing training received in the classroom. To enroll, there are certain procedures available on line. Following is a listing of some agencies and facilities that offer training, including a synopsis of the classes offered and Websites with additional information.

The National Fire Academy

The NFA is located in Emitsburg, Maryland. This training facility offers a myriad of training classes, and the cost is borne by the federal government. The facility offers housing and meals, and transportation costs may also be covered. The NFA has excellent instructors and offers courses of varying lengths and in various responder disciplines. Classes include response to different scenarios, incident management and specialized course offerings. This resource is sadly under-utilized by the responder community, even by the fire service. Most firefighters will never take advantage of the training and educational opportunities available here. The NFA also allows some of its training programs to be taught off-campus in every state in America. Many students will take classes prepared and updated by staff at the NFA at local training facilities. This type of training allows for standardized training to be presented by qualified instructors locally, in cases where the students are unable or unwilling to attend the training in Maryland. The NFA is available on line at http://www.usfa.dhs.gov/nfa/.

TEEX

Many training courses can be brought directly to your agency, with the cost of this training borne by the federal government. One such vendor is Texas A & M University Engineering Extension (TEEX). Its course registry is available at https://www.teex.com. It can provide training in a number of areas, and the staff and materials will be delivered to your agency or to a facility of your choosing. TEEX offers off-campus training in terrorism response, incident management and related fields. The instructors are very well qualified and experienced, and their ability to bring the training to you has added benefits, such as allowing for more of your personnel to be trained. TEEX offers a wide offering of courses from firefighting, incident management and similar classes at its site in Texas as well. Some of the training given in Texas is subject to reimbursement of covered expenses by federal authorities.

FLETC

The Federal Law Enforcement Training Center (FLETC, available at www.fletc.gov/osl) is another training agency, which will bring the trainers and equipment to your location, free of charge. They present a curriculum similar to TEEX. In the area of radiological detection and response, the Department of Energy Counter Terrorism Operational Support (CTOS), at the Nevada Test site will also bring qualified experienced instructors to your location. This program will also bring radiological detection equipment to the training, and offers three classes rolled into a one-week program. They will offer a one-day awareness program, a two-day operations course as well as a train-the–trainer (TTT) program. The operations class allows for students to get hands-on training with the detection equipment provided by the instructors. Successful students may be qualified as instructors if they pass the TTT program, which then allows them to present the awareness class at later dates. CTOS will provide the registration paperwork, student manuals and instructor materials to the successful TTT students. Students who are hazardous materials technicians, and who have completed the operations class are able to apply to go to the Nevada Test site for additional technician level training, as previously mentioned in this chapter. Again, the federal government covers all applicable costs.

Explosives Training

The New Mexico Institute of Mining and Technology (National Energetic Materials Research and Testing Center) (NMIMT) is a training facility, which requires the students to come to their location. The training at this facility includes classroom sessions and observations of controlled explosive detonations. As with the CTOS offering, successful completion of the program allows students to present the basic awareness class to other responders. The costs of transportation, rental car, lodging and meal reimbursement are all borne by the federal government. The class will explore various types of explosive incidents ranging from small letter bombs, to large VBIED scenarios. Participants must be registered through their individual state point of contact. This facility's prerequisites and course descriptions are available at www.respond.emertc.nmt.edu/. This training can provide invaluable insight into the world of explosives and their uses in the world.

WMD Training at CDP

The Center for Domestic Preparedness (CDP) offers students hands-on specialized training in Weapons of Mass Destruction (WMD) remediation. They will also present students with live-agent scenarios with respect to chemical agents. This facility is located at the US Army Chemical School, Fort McClellan, Alabama. Registration is the same as for CTOS and EMRTC, through your state point of contact. This training is available at http://cdp.dhs.gov/index.html. Again, the training here is very insightful, and allows students the ability to picture real-world responses to CWA incidents.

LSU Training

Louisiana State University (LSU) (National Center for Biochemical Research and Training, NCBRT) offers students training in biological incidents, and is geared mostly toward law enforcement response. They are available at http://ncbrt.lsu.edu/. The Rural Domestic Preparedness Consortium (RDPC) is led by Eastern Kentucky University will deliver training in the specific areas faced by responders in rural areas. This training group is available at http://www.ruraltraining.org/. Lastly, first responders may seek specific training opportunities presented by the Department of Homeland Security (DHS) by exploring their training catalogs. This is available at http://www.firstrespondertraining.gov/TEI/tei.do?a=home.

Getting Responders On-Board

The simple task of identifying the areas and responders in need of training is not the end of the process. Employers, legislators and the responders themselves need to understand the thought processes involved in recommending new areas of training for each responder group. There are some training classes which may be open to law enforcement only, and may offer materials which may be classified, or have law enforcement only restrictions. This would prevent civilians from participating in these classes, but there are certainly other classes which will gladly welcome civilians. The employers, especially in the private sector need to recognize the importance, and the benefits of expanding training options for their employees. For them to authorize training of their employees, they will need a comprehensive and convincing explanation of why the training is needed, and how it can be acquired. If the employers are not sold on the training, they may not agree to their employees being put in the recommended classes. Employees, especially new hires will accept the cross-training as a condition of their employment. The harder sell will be for the experienced, veterans, who will feel that there is no need for additional training. The convincing of the employer, or in the public realm, the administrators of the need will make the training easier to provide to these veterans. Again, the reluctant students need to have the importance of the training explained to them, so that they will at least be willing to attend the training with an open mind. The last issue to be addressed with administrators and employers is payment of employees for the training. For volunteer responders, there will be no cost. But for career professionals, if there is no training schedule available, overtime or compensatory time may be required. However, there is usually a mechanism for career responders to be provided with mandated, in-service training. The difficult area may be in asking private employers to pay the hourly rate for their employees, especially in an area which is new to the employee. This will be of paramount importance, since the private sector presents many different options for the professional responder to request, or in some cases, to debrief during an incident. These private employees are an excellent resource for incident managers, and employers need to understand what these civilian responders can provide to incident managers. Once they see the huge benefit of having trained personnel working for them, they will much more freely volunteer to have their employees trained in the recommended areas.

New Training Missions and Grants

In certain cases, additional training may lead to a new a new training or response mission for an agency. Should a new tactical mission be desired, there might be grant opportunities for the agency to qualify for not only for training, but also for equipment and other materials to support this new mission. In the current response environment, we have discussed the need for administrators and responders to do more with less. If doing more includes developing of a new mission, or responding to new types of calls for service, the opportunity to apply for federal grants will present itself. If we are required to, or decide to expand our current operational areas, there will be a need for new training and equipment. Explaining in the body or narrative area of a grant application that this is a new mission, or area of responsibility may result in funding for this new mission. While there are no guarantees, lack of local funds to accomplish a new response mission are generally favorably received by those charged with the responsibility of awarding grant funds.

When a new mission is explored, and grants will be applied for, it is imperative that equipment requested be covered under the grant parameters. Most grant applications contain guidance materials, which are available either online, or as part of the application. Grants are sometimes very specific in what materials, training and equipment may be purchased with grant funds. This information needs to be reviewed prior to submitting the grant application. Also, applying agencies need to be aware that maintenance costs and perishable supplies may not be covered under the grant. Applying entities need to be aware of this so as to ensure that their agency can plan for and cover related maintenance expenses if they are awarded a grant. For example, many of the Homeland Security Grants I have seen will not allow for the purchase of firearms. While the grants have allowed for many other items, firearms have not been a covered item. Accordingly, a grant narrative which asks for funds that will be allocated for firearms would automatically be denied for requesting an un-covered item. Those who will author the grants must study the helpline portion of the grant, and the parameters must be fully understood prior to final submission. Certain grants will have listings of which items are covered, and which cannot be requested. Even with a new mission, not all equipment or material may be covered under a specific grant program.

Grant Narrative Tips

Within the narrative of a grant application requiring new or replacement equipment and gear, certain language will greatly assist your chances of successfully obtaining grant assistance. For example, whenever a new mission is introduced, body of the grant should delve deeply into existing state and federal standards that are applicable to the new mission. Exploring what is required of the agency, and what training and equipment will be needed must be documented. If the grant is designed to supplement or replace existing materials, the narrative should explain that the grant is requested to assure 100% compliance with existing federal standards. For example, if a fire department is requesting a grant for structural turnout firefighting ensembles, it would be beneficial to note that if the grant is awarded, the department will be outfitted with 100% *NFPA compliant* turnout gear. In this case, the standard quoted would be NFPA 1971, which details standards for structural fire-fighting ensembles. This type of language will keep your grant application in the running longer, and with better results than if no standards or compliance issues are addressed. Any grant request that will ensure complete compliance with any standard or regulation will be given serious consideration by the agency overseeing the grants.

General Grant Application Tips

Whenever grant applications are submitted, it is imperative that they are reviewed for completeness and accuracy. Many grants are denied simply due to them being completed improperly. Directions contained in the grant application must be closely and exactly followed. In some cases, the grants to be awarded require a *match* from the requesting entity. What this means that is the grant will cover a certain percentage of the cost of the items, or training requested. The balances of the funds are required to come from the entity which was awarded the grant. Sometimes the match is as little as 5 percent, and sometimes it can be 30 percent, or more. There are also grants, which will cover 100 percent of the requested items and training. It is imperative that those who are requesting grants are aware of all of the rules, parameters and obligations prior to submission. Also, most grants will subject the agency receiving them to future audits. These audits are designed to ensure that the money requested is being used by the agency for the items, and mission it requested them for. My law enforcement and fire departments both were awarded federal grants, and we were subjected to audits on a number of occasions. When looking for grant opportunities, local and state governments may rely on each other for available information. Often, Offices of Emergency Management will receive information regarding application periods for certain grants. They will then forward the information and application deadlines to the interested or applicable parties, either via e-mail or regular mail delivery. Most of the grants will be advertised in advance of the application period, and will be followed by an announcement by the agency controlling the grant, once the application period begins. If your agency is interested in locating grant opportunities online, any decent search engine will provide guidance and potential grant opportunities. Simply enter key words such as federal grants, law enforcement grants, fire grants, training grants etc., and the search engine will generate a list. It may take time to sort out the grants to see which ones you are eligible for, but should you secure a grant, it will have been worth the time. Searches can be refined, or narrowed down if you are getting too much information, or information which is not applicable to you or your agency or the mission for which it is charged.

Stress Safety in Your Grant Application

Another key area to address in the narrative portion of grant applications is that of safety. Safety of responders and civilians has long been an area of deep interest for those who administer grant programs. If your agency has a genuine need for materials and equipment, or training whose purpose is to ensure safety of responders and the public, be sure to stress that aspect of the grant in your narrative. Many responder communities have extremely dangerous jobs, and administrators and employers are always seeking ways to enhance the safety of these responders. Manufacturers also recognize the need for safety, and research and development of many different types of products are always being pursued. Manufacturers will also advise the proper response entities if their products are available under certain grants. Trade journals are an excellent source of information on new products and technologies that may be subject to certain grant programs. Technology is constantly being developed, which is designed to assist us in being safer as responders. For police, body armor is constantly being reviewed, and newer, lighter, more durable products are being offered. For firefighters, technology on firefighter safety while working in burning buildings is always being explored. Systems that will track the location of firefighters when they are in need of assistance or those that measure temperature of buildings and remaining air in SCBA are examples. Certain companies are developing new technology to assist hazardous materials teams in product identification. When these technologies are brand new, they are often very expensive, and out of reach of many agencies. Progressive, forward thinking departments and agencies are always seeking these products and technologies to make their responders safer and allow for better operations. However, due to costs, they are sometimes unattainable by even the most progressive responder agencies. This makes grants a viable alternative to securing these products and services. For example, under the assistance to Firefighters Grant, one of the areas, which will be funded, is operations and safety. This is a broad area for agencies to apply for funding, but this also allows for a broader base of products to be applied for. Firefighter *bailout kits*, for example are an example of a tool used for safety. These kits contain a load bearing rope, a descending tool, and an anchoring point of one type or another and a harness. The purpose of the kit is to allow firefighters who may be trapped on upper floors of a house or apartment building or other structure to deploy the anchoring device, drop the rope out the window, attach the other end of the rope to the harness, and use the descending control device to control the speed at which the firefighter will lower themselves to safety. These kits came into prominence in January of 2005 after six New York City firefighters were forced to jump out of a sixth floor window onto the ground during a fire, killing two and seriously injuring four others.[1] The next grant year, many departments sought to purchase these bailout kits to enhance the safety of their members. Many applicants were successful in arguing that there was a genuine need to equip their firefighters with this potentially life-saving equipment. Another example of newer technology for firefighters would be the recent influx of equipment used to track the location of firefighters at incidents. Firefighting is an inherently dangerous business, and the hazards in structure fires are often unseen until it is too late. Structural collapses often lead to injured or trapped fire crews. There are many companies that offer equipment designed to be worn by firefighters, that will transmit a signal, and with the receiver, rescue crews can locate the signal transmitted by the injured or trapped firefighter. Regardless of which type of equipment is being sought, stressing safety of responders in the narrative of your application is always beneficial.

Research Your Proposal Prior to Submission

The obvious goal for any agency when applying for grants is to be awarded a grant in the specific area for which the material was sought. Many times, other agencies have already applied for, and received grants very similar to the one that your agency may be pursuing. It is always a good idea to research other agencies that have been awarded a grant in this area and then to review their applications. Often, similar wording may be beneficial in your application for a grant. Some grant programs will post previous successful applications for review during the period for which the grant is accepting applications. Fire Act program administrators offer free seminars for the fire departments planning to submit applications to attend. These seminars will review the applications, point out areas where others have succeeded, and more importantly, areas where others have failed. These types of free seminars are very helpful in preparing for the submission of your application. As the saying goes, don't reinvent the wheel. Many agencies will be more than happy to discuss their strategies for successful grant applications if you call and ask for assistance. Take advantage of the work done by others, especially if they were successful, and model your application after the successful ones.

Use Regional Program Grant Applications

Also, given the battles for funding dollars, there is another strategy, which may yield better results. This strategy involves making an application for funding for a regional program or project. The benefits of regional grant applications are that more than one agency or jurisdiction will reap the benefits of the grant. This type of award allows for better spending of the grant money, which is attractive to the program administrators. For example, if a single agency makes a request for a SWAT vehicle, and several related pieces of equipment, the application is graded on the benefit of the vehicle and equipment for the population served. Conversely, if a group of towns, or even a county were to make the same application, and in the narrative explain that the grant is a regional application, it will be reviewed and considered more closely. The benefit here would be that the same federal dollars will be allotted, but the population served, or which will benefit from the funding is much larger. All of the towns that would be partners in this type of application would need to have agreements in place regarding staffing, training, housing the vehicle and equipment, and dividing the matching part of the grant, if one exists. Many of the current homeland security grants and funding are shifting to this type of regional project, and your chances of securing some of the precious federal dollars earmarked for grant funding will be greatly enhanced if your application is for a regional project. You have a better chance applying regionally than if you prepare and submit the same application, requesting the same materials alone.

Lastly, some tips and suggestions for grant applications have been reviewed. These are just some ideas and tips, which you may find helpful when applying for grants. There are so many different types of grants and grant programs that to list them all would not be feasible. You may choose a search engine of your choice, and enter key words such as grants, federal grants, matching grants, or grant programs. There will be many sites to which you may be referred, and at this point, simply refine your search to limit the sites you may find useful. Always remember to look for application opening and deadline dates, prerequisites and parameters. As already indicated, look for help information which may be associated with the grant applications as well.

Summary

In summary, we have covered a great many topics in this chapter. We have reviewed the importance of choosing the proper, credentialed and qualified trainer or instructor. This person must be familiar with your policies, procedures, capabilities and limitations. The trainer must be able to relate how this new training area will be of benefit to your staff. These trainers must have the ability to handle disruptive or reluctant students and should be experienced in the field that they will instruct. Sometimes, it will be beneficial for the instructor to be a member of your agency, and sometimes, it would be better if they were from another outside agency.

The locations for conducting training were also reviewed. Often, it is easier to conduct training at your facility or training location, while sometimes it is better to conduct training in a neutral location. Training facilities or academies which are dedicated for training are also an excellent option as a training locale. At these dedicated facilities, the instructors can be provided for you, and therefore, the need for your own staff members to serve as instructors is not necessary. Further, the staff at the facility should have extensive expertise in the field, which will be addressed during the training.

Trainers and students need to have the same expectations regarding the training, which is to be presented. If one or the other group has drastically different expectations, the training will not be as smooth and successful as it could be. The student and instructor need to have similar agendas for the materials to be properly presented and received. Also, training providers were discussed, including some that require students to go to the trainer, as well as trainers that will come to your agency. Federal funds may be used to offset the expenses if you must travel, and if the materials are coming to you, the cost may be borne by the federal government. Several of these vendors and contractors were listed, along with synopses and web site addresses. Many of these course offerings have limited class sizes, and some will require previous training, which was conducted at a lower level than will be offered at the off-site class. These contracted training agencies have extremely well-qualified, experienced instructors, and they offer tremendous insight into the areas in which they will train. Some of the training classes will offer the ability for students to return to their facilities as instructors. This truly enhances the training experience, since agencies may have a new instructor on their staff who can re-present the material to other members of your staff.

The last area of this chapter dealt with grant opportunities, and tips for your grant application. Today, there are many different grant opportunities, some of which are responder specific, such as the Fire Act, and others that may be applied for by any agencies. Correctness on the application is among the most critical of factors, as those reviewing the application will quickly discard an improperly completed application. Prior to making your application, you must be certain that you are eligible for the grant, and that the money or equipment you seek is covered by the grant. Deadlines for the starting and ending periods of the grant must be closely adhered to as well. A request for funding an area that is not covered under the grant will also result in a quick rejection of the application. The narrative must be concise, thorough and well written. Essential components of a successful narrative will focus on safety issues, as well as the funding bringing the agency into 100% compliance with existing standards or regulations. If your agency is expanding its role into a new mission, grants are an excellent manner in which to subsidize this new response commitment. As already noted, careful attention must be paid to the equipment required in this new response area, to ensure the proper grant application is filed. Regional grants are also a hot-topic in the grant community. Federal dollars that will be spread out over a larger area, and affect a larger population will receive high marks by reviewing authorities. Grant writers may refer to successful grants that were authored by other agencies as a means to increase the likelihood of their agency being successful in receiving the grant. Seminars and help pages, which are offered online, are also an excellent resource for your grant writers. A final thought regarding grant opportunities and grants themselves should be noted. Often, the agency that offers the grants will only make certain items available. While grant equipment is always welcome, the process of offering certain items seems wrong. Agencies should have far greater input as to what they need funding for. For example, the receipt of equipment is always welcome. New equipment requires training, which may generate overtime issues. Generally, grant funding will not allow for the payment of overtime for the requisite training. Eventually, we will have equipment in place, but no responders trained or competent in its use. This is a serious issue facing administrators soliciting grant equipment. Equipment that remains in storage during an incident is useless. Funding needs to be made available to train responders, even if it generates overtime. Obviously, this is aimed more at career departments rather than volunteers. It is extremely unlikely that grant funding will pay a days wages for volunteers to attend, but as previously discussed, volunteer training can be accomplished in a number of ways.

There have been several areas where responders can be cross-trained, which have been explored herein. Funding for the training will always be a touchy subject, especially in the private sector. We need to ensure that those responsible for the decision-making process are well informed, not only on the need for training, but the available resources to provide it. Grant opportunities and federal training partners can ease the financial burden, but a commitment needs to be made by the agencies and their administrators as well. There are many training options available to the response community, and there are excellent instructors willing and able to present the materials to us. We are all facing shrinking budgets and escalating expectations. These two issues presented to us simultaneously are very serious and require creative thinking and solutions. We must factor in all other workplace issues which we face when deciding on how, when, where, and on which topics to train. Training classes and requirements change very rapidly for responders, and we need to keep up on the latest versions of these classes. Responder-specific problems remain, even as we try to remain vigilant across all realms of response and incident prevention. Drug activity, quality of life crimes, false alarms, EMS calls for service, such as people with colds and the flu will always remain at the forefront of our responders' activities. While these are certainly service areas that need to be dealt with, there are other areas that need to be considered as well. Complacency with regards to training can have dire consequences for not only the citizens we serve, but for the responders themselves. Administrators must find the creative solutions to maintain a level of service that the public expects, coupled with allowing our responders to be as well trained as possible. Failure to address the training needs of our responders will not be tolerated should an incident where the cross-training would have been a difference-maker is encountered. America's response community was given a "pass" on September 11, 2001, but the public will not be so forgiving if we experience another large-scale incident. The problem has been identified, and solutions have been proposed. All we need to do now is act on the solutions presented.

Endnotes:

1. "3 Indicted in Deaths of 2 Firefighters Who Jumped." Article published March 30, 2006. Available at: www.nytimes.com/2006/03/30/nyregion/30fire.html.

Chapter 11

Online Training and Distance Learning: Friend or Foe?

The last chapter of this text dealt with where to get and pay for training as well as grants and grant application tips. One area conspicuously omitted from that chapter is online and distance learning opportunities. Online training may be available for a number of training topics and is, in fact, training. The relatively new concept of distance learning is not training, but rather education. Distance learning is where a student may take classes toward a degree, either associate, bachelor level, or even graduate course offerings, but from home rather than on-campus at a college or university. These distance-learning opportunities present themselves as courses for collegiate credit, and are taken from the home of the student, while online. Successful completion of the courses results in the awarding of college credits, and when sufficient credits are amassed, a diploma is awarded. Many of America's responders have seen this as a viable method for continuing their formal education, and for preparing them for possible promotional opportunities. Education is never a bad idea, and people who choose to continue their education via these programs are to be commended. Given the time constraints, family commitments, and costs associated with enrolling in and completing college classes, the distance-learning option is very appealing to a great many student responders. The trend of distance learning is gaining momentum, as are the phenomena of online training, or virtual academies, which allow training at a computer station. But are these good ideas for our responders?

Online training

The first area we will explore regarding this issue is the trend toward online training for first responders. Some states refer to the online training classes and courses as a virtual academy. In many ways, it is a virtual academy, but in many others, it is without substantial merit. In fact, I have spoken with a great many student graduates of virtual academies, and I will agree to call them virtual. However, in my opinion, the students gain virtually nothing from these classes and class offerings. Under the doctrines of NIMS, and other executive orders, which vary from state to state, certain training requirements have been mandated and outlined. Given the financial inability to cover overtime costs and to pay instructors to come to teach the required classes, the concept of providing responders with training courses online was devised. While I don't really have an issue with the intent of the program, the application has a lot to be desired. Several components of this online or virtual training will be discussed throughout this chapter, leaving the reader to decide whether or not the training strategies endorsed by these programs are effective.

Virtual Academies—Helpful or Harmful?

Virtual academies and online training offerings in different response areas are wide-ranging and varied. The goal is to provide the course content to the responder in a timely and cost-effective manner. In the coming pages, some of the courses offered in these virtual academies will be explored and dissected, providing an opportunity for the reader to compare the intended training goal with the reality. The first training classes to be discussed are incident management classes; specifically, ICS 100 and ICS 200 which are available online. I will begin with these classes since they are among the classes that I instruct on a regular basis. Instructing some of the materials in these classes is rather difficult, since it is very technical and dry material.

Police, fire, and EMS personnel are all required to have incident command training at some level because they may be required to fill the critical position of incident commander (IC). Within these courses, there are terminology issues, position titles, and other components that students find mundane, and boring. Anyone who has taught or taken incident command classes would probably agree that applying the materials in each unit to a prescribed scenario heavily drives and reinforces the course materials. The trouble with the virtual academy offering is that there is no true mechanism for testing the student's ability to apply the course material to a scenario. In presenting the course materials to my students, I have noted their ability to parrot, or regurgitate the information presented in the classroom. For example, after being instructed, they can tell me that the general staff consists of the planning section chief, logistics section chief, operations section chief and the finance/administration section chief. They are also able to describe the duties and responsibilities of each of the section chiefs. Once they feel confident in their understanding of the roles and responsibilities of each section chief, a practical scenario is presented which the students must manage, and apply the tenets of incident management.

In almost every group, in almost every class, I have observed that the students fail to apply the information, even though they can explain the roles of those involved. After allowing each group to draw and fill out an organizational chart of the incident command structure, I usually must present the material again, referencing the scenario in order to allow the students to see a properly completed organizational chart. Common errors by students include assigning the same person to more than one functional area, confusing the concepts of divisions and groups, continuing issues with position titles, and making the command structure larger than it needs to be. Once the material is reviewed using the organizational charts provided by the students, the class makes noticeable progress. The students are then able to make the link between the classroom and the scenarios.

Should the same students take the online version of this class, there is no mechanism for presenting a scenario to be completed by the student. Additionally, there is no way for the students to see whether or not they are correctly applying the information contained in the unit to the scenario. Consider also the fact that there is no mechanism enabling students to ask questions. If there is no forum for answering questions, what do students do when they are confused about a concept or block of instruction? In a classroom setting, they would simply raise their hands and ask the instructor to answer or clarify their questions. Using the online version of the class, they have no way to get an answer, unless they know a qualified instructor. As a former student, if you had questions about course materials which went unanswered, how confident would you be regarding the class materials? You would probably be very leery about the subject matter and hesitant to apply it in a real-world event.

The next issue of course is the final exam required for course completion and credit. If the class being taken is mandatory for your entire staff, I would be willing to wager that the answers to the test are readily available for the students to enter in the computer. I would be curious to know the pass-to-fail ratio of online courses. Even given the fact that the tests can be taken as open book exams, the failure rate is probably in the low single digits. I also wonder if the time each student takes to complete the exam is recorded. Similarly, does the virtual academy log the amount of time that students are logged into the system during the time the materials are being read? For the time online regarding taking of the exam, I would guess that the time is only slightly longer than is required to mark each answer. Some students probably never even read the test questions, and why should they if they have access to the correct answers? But, cynicism aside, what if a student earnestly takes the class, and then takes the exam. Should the student pass, the exam is not returned, nor are the correct and incorrect answers reviewed and explained. If seventy percent is passing, and you get your e-mail notification congratulating you on successfully completing the class, what was your score? If you do not know what you did wrong, you may be reinforcing incorrect information, believing it to be correct. If this is the case, what information did you really glean from the training?

In the case of training which is scenario driven, such as incident command training, virtual academies are certainly a foe. There needs to be a mechanism for scenarios, and students must be able to ask questions. Without these two critical factors, students will not learn all that they should regarding incident command. If it is not done right, why should departments bother doing it at all? We do it because we want to present impressive numbers of those who are trained to our superiors. These are certainly empty numbers.

Specific Examples Where Online Classes Are Less Desirable

There is a class offering which has certain incident command characteristics, but which is a freestanding offering. This class is the National Incident Management System I–700 (NIMS) and is one of the core classes required for all responders in America today. NIMS compliance is required for states to qualify for and receive Homeland Security Grant funding. When presented in the classroom, NIMS takes approximately four hours. The material is very involved, and includes a number of acronyms and concepts that can confuse students. The exam at the end of the block of instruction is difficult and, even when taken as part of an open book protocol, requires significant time to complete. In speaking confidentially with a number of responders, they admitted they took the NIMS class online, and went directly to the test since the answers were made available to them by colleagues. Because NIMS compliance is required for federal grant funding, administrators directed their staffs to complete the class online as soon as possible. As with most other courses offered online as well as in the classroom, the ability to have a live instructor with experience and background in the areas covered by NIMS is ideal; online instruction pales in comparison. In fact, several opinion pieces have been published in various responder trade journals debating the need for the NIMS training and some of the doctrines contained therein. In reviewing these op-ed pieces, it is apparent that there is still a great deal of confusion regarding some of the concepts and principles of the NIMS class. Having a qualified instructor present the materials would be a tremendous asset to an agency and the ability to field and answer questions would put to rest much of the confusion surrounding NIMS. Again, the benefits of a live instructor outweigh the convenience of online instruction.

There are other class offerings, which vary from state to state. One of the virtual academy classes discussed next is counter-terrorism awareness or a similar type class. This topic is another area where I have been credentialed to teach in a classroom setting. As with incident command classes, I have presented this material on many occasions, and the version that I teach takes two full class days. This type of course is broken down into different modules or units, each with a specific function or goal. As with the incident command classes, some of the modules are very dry. Conversely, some are very detailed and deal with history, attitudes, criminal enterprises, and warning signs, which may indicate the presence of terrorist activity. Some of the materials such as document authentication are very tedious, and subtle nuances need to be addressed. In presenting this class, many questions were asked, which generated debate and discussion among class members. As with other online classes, there is no mechanism for asking questions for this block of instruction. When classes deal with sensitive or detailed information, students are bound to have questions. If there is no instructor to ask, the student does not get the full benefit of the class. Also, given the sheer size of the student manual and the fact that it takes two class days, reading this material for that amount of time is going to generate issues for students. Furthermore, responders are not used to being tied into a chair for an hour or two, let alone two days. After a certain point, they may be reading the words, but they are not retaining or grasping the author's concepts. As noted earlier, if the class is being taken because it is required rather than because of the student's desire, a two-day class such as this is doomed. The student will reap no benefit. Again, the overwhelming odds are that the test answers will be provided with the result that the student gains no real benefit. Left to their own devices with tedious mandated training, many responders may "shrink" out of the training window and "browse" elsewhere for the duration of their required online stay. Some responders may even log in, shrink the class materials, complete other duties, and log out after a prescribed number of hours. If there is no way to track page movement or activity on the web-page, getting around this type of training would be easily accomplished. Again, if there is no demonstrated desire on the part of the student, there will be virtually no benefit.

Hazardous materials and CBRNE response classes are also offered on some online training sites. Again, these courses are generally mandated, so there is a need to demonstrate that the training has been conducted. Depending on the response level to completion requirements, these courses can range from four to sixteen hours, or longer. The field of hazardous materials leaves very little margin for error when the principles are applied on the street. If the online academy is to serve as a vehicle for refresher training, the material must be absorbed by the responders. Hazardous materials are around us every day, and the potential for a release of these materials is real. Refresher training generally covers the use of the North American Emergency Response Guidebook (NAERG), and a review of certain principles of chemistry and physics. Other than hazmat teams, first responders are trained in defensive actions only and offensive incident mitigating techniques are left for hazmat teams. Failure on the part of responders to familiarize themselves with these principles can have obvious deadly consequences. As with other online offerings, much of the hazardous materials class is very dry, and reasonably technical in nature. Often, responders feel they have a sound understanding of this type of material, but hope to never need to utilize it.

With regards to CBRNE and CBRNE response, many of the same concepts apply. The ability of responders to recognize potential indicators of terrorist attack will be critical in saving lives of victims as well as other responders, and lessening the impact of the attack. Signs including the previously mentioned odors must be readily recognized, and the responder must be ready to act upon witnessing or detecting them. Since the hazardous materials and or CBRNE response training will probably be mandatory, the students will probably not be highly motivated. Often, if the training is set up as refresher, there is no written exam to test the student's recall. Furthermore, the lack of an exam will make the responder even less likely to truly read and attempt to grasp the material. And, as previously stated, there is no mechanism for scenario-based learning or for asking questions. As with the other classes covered so far, there is virtually no benefit in presenting either of these courses on line, especially as refresher training.

Online Training Which Can Be Helpful

Thus far, we have addressed online offerings, which may be considered mandatory training. There are, however, other training opportunities offered online which are not mandated. The National Fire Academy (NFA) developed a series of incident management scenarios designed for fire department personnel. Years ago, I had a part-time job as a per-diem fire lieutenant. The chief of this fire department was very proactive in training his personnel. Once he learned that the NFA had developed and released this program, it was added to our training library. This series simulated fires and responses in five different scenario locations. The program required you to respond to the incidents and make real-time incident management decisions following prompts. This series required that tactical decisions be made and, if the wrong decision was made, students were prompted to reconsider their choices. Since this was a NFA offering, some tactical discrepancies may have been based on regional practices. While there was no forum for the student to ask questions, the program was reasonable and realistic. Students were given initial dispatch information, a list of resources and crew sizes, arrival times and locations, etc. Based on this information, students made tactical decisions such as hose line size, points of entry and assignments. Unlike incident command lessons online, this type of scenario training was beneficial. Many of the decisions made were in line with the thought processes involved in incident management tactics at the time. Since the training was not mandated outside the fire station, it was generally well received. In fact, the other members of my crew made a competition out of the training, and we tracked the high scores on each of the five scenarios. Accordingly, this is a case where online type training can be implemented, and will be beneficial to the student. The training was fun, realistic to a certain degree and completion was only required in-house. These types of classes will work as online offerings.

The Federal Emergency Management Agency (FEMA) also offers a large number of classes covering most responders. Their online training is referred to as Independent Study (IS). When a course from the Emergency Management Institute (EMI) is completed successfully, an e-mail containing a link is sent enabling students to print their certificates. As with most everything in today's world, there is good and bad in this system as well. The usual bad aspects are present. If the student has issues, there is no mechanism in place for asking questions so an instructor credentialed to teach that particular subject must be located. Also, this training site offers classes on incident command, which we have already discussed. Furthermore, the same issues regarding scenarios and inability to review completed exams are present in these versions of the incident command classes.

FEMA offers a series on emergency management, blocks on NIMS, incident command, radiological awareness and response, OEM classes, courses on volunteers, weather, and WMDs, etc. There are dozens of classes for the self-motivated student to dive into. Most are well designed, and require a test for the student to receive "credit" and a certificate. In many of the courses, quiz questions pop-up in the middle of each lesson. These require correct answers before the student is able to proceed further with in the lesson. While the site offers some courses that are also offered on other sites, most—if not all of the offerings—are not *required* by any particular agency. Accordingly, visitors to this site are often people with a desire to educate themselves in one or more areas of response. There are enough course offerings in multiple areas to satisfy almost any responder who desires to learn. Groups, such as politicians who are required to take classes such as NIMS, may also utilize this site for mandated training. In cases where the training is not mandated, a site such as this one has many upsides, with limited down sides. The other advantage to sites such as these, and even virtual academies, is that they are available twenty-four hours per day. The sites are also free to the user, although state training academies may require registration, and verification of employment. As a training tool, these virtual academies offer a wide array of classes, but the students must desire the training and must be willing to be tested. The resource is a good idea, but it needs to be better managed and tracked before it can be deemed to be an effective and realistic training environment for all of our responders.

Distance Learning

The concept of distance learning is very different from that of the virtual academies, which are very prevalent in today's responder communities. Many responders are anxious to continue their formal education, but are confronted with a great number of obstacles. These obstacles include hectic, often changing work schedules; overtime—both mandatory and voluntary; family commitments; time constraints; financial considerations; and logistical concerns. The first responder today, as with most working Americans, has extremely demanding work and personal schedules. Often, there is simply not enough time in the day to complete each and every task scheduled for the day. To add the complexities of a school or educational experience to this schedule is very difficult. Even with research studies recommending against rotating shifts, many responder organizations still utilize this practice. Firefighters may work many different schedules, some of which require a twenty-four hour work day followed by either forty-eight or seventy-two hours off. They may work a fourteen-hour shift, or a ten-hour shift, and some departments work an eight-hour shift, Monday thru Friday. EMS and hazardous materials team members often are faced with rotating schedules as well. Law enforcement personnel also work a variety of shifts, some ten hours, some ten and one-half, some twelve, some eight. Many of these shifts are steady, but some require rotation. The ability of these potential students to enroll in a traditional institute of higher education is severely hampered by these scheduling issues. Making matters worse is the overtime which responders are either forced to work, or which they schedule themselves to enhance their salaries. Firefighters cannot just walk away from a fire because their shift has ended. Similarly, EMS workers may be dispatched to an emergency medical call requiring transport near the end of their shifts. Police officers may be stuck on the road at a service call, accident, or may be processing arrest reports when they are scheduled to report off duty. They are also required to attend municipal and state court hearings that may be scheduled for their days off. When duty calls, the responders must be ready.

Even if their work schedule is such that classes may be scheduled, there are family issues that need to be addressed as well. Many responders with children would rather spend some quality time with their children, or attend sporting and related after school activities with their children. Given the choice to attend school or to spend time with their families, most responders choose family. Adding to these obstacles to continuing education is the expense. The cost of college today is so high that many potential students are unable to afford the tuition. For this reason, financial aid is a viable option. However, working adults who are carrying mortgages and sometimes a second mortgage, car payments, and the costs associated with raising children, another bill is a major deterrent. Simply put, college is cost prohibitive for working responders, especially those with families or a spouse who does not work. These potential obstacles, in addition to logistical issues such as travel time, the distance required to attend school, and work required outside the classroom all serve to deter responders from continuing their education.

Benefits of Distance Learning

Professional responders see the potential benefits of bachelor's degrees, and even master's and doctoral degrees and employers should be supportive of any employee who seeks to become better educated. Sadly, the practice of reimbursing employees for expenses related to continuing education is being phased out for financial reason. In times past, employers may have also considered allowing employees to attend classes while on "company time," but this practice is also gradually disappearing. Employers today simply do not see the result being worth the expense when overtime costs are considered. Nor do they appreciate how the degree will enhance the employee's job performance. Rank-and-file responders envision college degrees and advanced degrees as vehicles leading toward possible promotions. Also, employees nearing retirement see the college degree as providing them with a marketable skill to be used after retirement. Education is a good thing, and continuing education is always a worthwhile endeavor. The sad fact remains that for most responders, there are too many obstacles to overcome in the pursuit of continuing the formal education process.

Fortunately, there are now options that may allow responders to continue their formal education without having to make some of the difficult choices referenced previously. Schools that recognize these issues have begun offering degree and advanced degree programs to potential students via the Internet. These programs have several positive aspects, but there are also some negative aspects as well. The programs will offer associate, bachelor, and master degrees, in limited fields. Since the course content is available online, many fees associated with physical attendance at schools can be waived. The classes are set so as to allow students to review the materials at their own pace. Scheduling issues are easily remedied when students can complete the required course content as time becomes available, rather than in a strict timeframe structure one would encounter on campus. In addition, these online classes are much cheaper than if the student attended the college in person. Students are able to e-mail work assignments and projects to their professors for evaluation and review, and this feature is of great benefit to the students.

Most colleges and universities that offer distance learning options have curricula including fire science and criminal justice, along with sociology in one form or another. This allows the responders to enroll in a course of study that is relevant to their careers. These schools sometimes give credit for life or work experiences in lieu of prior college credits. This feature too is very helpful in allowing the responders complete their courses in a timely fashion. Each school has its own review procedure for awarding credits for work histories, and not all will allow the substitution of work in a related field with classroom study. Therefore, responders wishing to substitute work history for academic credits should be familiarize themselves with the procedures of their chosen educational institution.

Potential Drawbacks to Distance Learning

While there are many appealing and helpful aspects to the distance learning, there are also some negative issues. The primary issue is their accreditation or lack of accreditation. A degree or certification from any school or college that is not accredited is of limited value. If the aim is only to continue an education and gather more information, this lack of accreditation will be of no matter. However, if the goal of the degree is career advancement or promotion, accreditation will be very important. A reviewing authority will want to be certain that an accredited institution provided the education, so that it may ensure that certain educational standards have been met. It would be a shame to invest the time, energy, and finances in a college program only to be told that the degree is not recognized. Accordingly, students must conduct research prior to enrolling in one of these distance-learning schools to ensure that proper recognition will be afforded to the achieved degree. This will also be true should students opt to pursue a master's degree.

Certification versus a Degree

The last area of online, or distance education to be addressed is in certification, rather than an academic degree. Many associations offer *certification* in certain areas or fields. I have seen, and been offered membership in organizations that will allow me to become certified in homeland security, for example. Certifications can be a tricky situation for those with little experience. The first step to consider in these situations is to inquire who will be providing the certification. Will this certification, much with aforementioned degrees, be recognized by your agency, or other agencies? In theory, I could offer a block of instruction in a few choice areas, and upon successful completion of my prescribed curriculum, I could certify you in homeland security, for example. If, however, you added my certificate to your resume, and presented it for review during an interview, what would be the result? Would any employer recognize me as an entity capable of course development and certifying you in any area? Prior to taking advantage of any such offers, it is imperative that research be conducted to ensure that any degree or certifications offered will be recognized.

There are legitimate organizations that offer a series of courses, and then certify successful students meeting prescribed criteria. Allow me to provide an example. Many years ago, New Jersey had a training organization under what is now the Division of Fire Safety (NJ DFS). This training entity was called the New Jersey State Fire College. The State Fire College provided various courses ranging from fire officer development, safety, engine company, and truck company classes to live fire training in many different forms. If a certain series of courses were taken and the appropriate forms were completed, the NJ State Fire College would *certify* you in a designated area. In my earlier days, I completed the block of instruction on truck company operations. I completed the required forms, and received my truck company certification. Since the certifying body was a state agency, the certification was recognized throughout New Jersey. I do not know if any other state would have recognized the certification, but it was certainly recognized in New Jersey.

Other Online Training

There are other organizations that will offer online training in blocks of classes. FEMA, via its independent study offers a series of classes on incident management, Office of Emergency Management, and other general fields. Certainly, any student who took a series of classes at this institute would be recognized as having a formal education. It is important to remember that certifications do not entitle those certified to any special privileges. The National Fire Academy is another example of an organization that may offer certifications to students. A certification endorsed by an institute such as the NFA would certainly be recognized by most state agencies. State authorities may also offer certifications, and again these are subject to acceptance or rejection by other entities. As a result of my studies and training, the NJ DFS has certified me in a number of areas. Since there are no current existing federal standards, as proposed by NIMS, I do not know how my training will be received in another state. Certifications are only good if the standards used to generate the certification are good as well. As long as your agency will accept either the training, or the certification, you will be all right, but someone needs to recognize your education.

Summary

In summary, this chapter has explored virtual academies, distance learning, and certifications, which are non-mandatory. At the beginning of the chapter, a question was asked regarding these types of learning. After reviewing the training types, has a decision been made? Are they friend, or foe? The answer will be different for each type of training and each reader. Organizations that require certain training be conducted online or at virtual academies need to be responsible regarding the course and course materials. In cases where the student is allowed to log into a session, shrink the screen, and continue regular work activities, a disservice is being committed. If materials or classes are required, it is incumbent upon the responder, as well as the employer or organization to assure that the materials are reviewed and the classes taken. Having one responder get the answers to test questions and either posting them or making them available to co-workers or other responders serves no purpose. A responder who submits the provided answers to the reviewing authority and then accepts a training certificate is a fraud. Anyone who takes and successfully completes any course should have learned something. To make no effort to review or digest the course materials is disingenuous. The responders, agencies, organizations, and the public deserve more from us. I do not believe that all responders who are required to take these types of classes will take advantage of these shortcuts, but no one should.

The subject of reluctant students was addressed earlier in this book, but the concept applies to these virtual academies. If responders are forced to complete classes, they will be getting no value from them in the computer rooms of our agencies. It would be far better to require the classes to be taken under the direction of a competent instructor. The goals of those who developed the idea of virtual classrooms for responders were pure, and were genuinely designed to allow responders to get training. However, the mission of agencies to get their people trained may have gotten off course. If this is true, administrators need to require their responders to review the materials, and take the courses as individuals. In cases where students take classes out of desire or a thirst for knowledge, this will not be necessary. Simply requiring personnel to take courses merely to indicate that we have trained a certain number of people is a hollow statistic if the classes are not actually completed. Instructor-driven classes make more sense if education is truly the goal.

Other issues as well were reviewed and discussed including the inability of students to ask questions in a virtual classroom. This area alone should make instructor-led classes the preferred delivery method. There must be a satisfactory process for inquisitive or confused students to get guidance and information about issues raised during training classes. Materials that are not fully understood leave students unsure of how to apply the material in the streets. What good is training if it is uncertain that the information was received and properly understood by a student who may need to apply it for real?

The last flaw in some training conducted online is the inability to apply the materials to scenarios during the class itself. This is most obvious in incident management classes, which are heavily driven by scenarios. Students can often parrot materials in an effort to demonstrate understanding. In these classes, it is also imperative that the information presented be applied to the system during scenario training. The lack of a system to allow for this seriously hinders the students from fully grasping, and feeling comfortable with the material. Accordingly, having no forum for scenario training or for asking questions during training can certainly make a case for this type of training to be considered a foe rather than a friend. Training which statute or departmental authority mandates is not always interesting or even appealing to the responders. However, it is mandated for one or more reasons. Endorsing a system that may not guarantee complete compliance realistically serves no legitimate function for either the student or the agency. Hazardous materials and CBRNE response classes fall into this category as well.

Also in this chapter, distance or online educational opportunities were discussed. These educational opportunities also have pros and cons. The pros are very compelling, and include financial savings, the ability to continue the educational process, the ability to digest and review the materials at the student's own pace and the certification or the awarding of degrees. These attributes make this type of learning extremely appealing, as does the possibility of using work history as a forum to receive collegiate credits. This learning system also does not force responders to choose between time spent with family and the desire to seek additional education. Further, making these courses even more attractive, they offer areas of instruction consistent with our job titles and goals. Overall, these programs are well designed and thought out, and are becoming increasingly popular in our society. The potential drawbacks include the possibility that the chosen institution may not be accredited. Accreditation will be very important if the courses are taken to enhance a resume or for promotional opportunities. Agencies requiring certain numbers of college credits or college degrees will certainly want to ensure that degrees put forward for consideration come from duly accredited institutions. Therefore, the distance learning phenomena is more of a friend to the responder community. Students must research the accreditation of the institution prior to deciding if and when to enroll. Once that step is taken and the student is satisfied with the answer, the educational process can continue and will ultimately prove beneficial to the student.

Finally, voluntary training at various online institutions was reviewed, especially in respect to seeking certifications. If the certification sought is a voluntary certification, the training is generally sound. Anyone seeking to take a course or series of non-mandatory courses is certainly a motivated and dedicated student and responder. Given the large amount of training that responders are required to participate in to remain qualified as responders, any non-mandated training sought out by any responder should be encouraged, and recognized by the agency. The self-motivated responders who solicit training sites on their own time, simply to expand their education would most likely not cut corners in taking the prescribed courses. Many agencies and organizations offer non-mandatory courses online and these entities will certify students upon successful course completion. Here again, the only potential drawback to this type of training is whether the required certification is recognized out of the organization or by other agencies or states. The recognition by other entities probably will not matter if the student takes the training for his or her own knowledge and benefit.

In the final analysis, each agency, each course offering and each student must be evaluated on its or their individual merit. Certainly, opportunities to circumvent the learning and educational processes exist. Serious-minded responders and organizations will not tolerate these shortcuts, and will take steps to ensure that course materials are reviewed, and that exams are taken without the benefit of answer sheets. Therefore, there is no real answer to the question of whether training options are friends or foes. In this area, it is the responder, not the educational forum who can be evaluated as a friend or a foe.

Chapter 12

Planning a Tabletop Exercise

At this point in this book, many training areas have been explored and certain courses and classes have been recommended. As a diligent forward thinking, proactive responder and manager, you are ready to take the next step. But what exactly is the next step, and how will it be undertaken? The logical progression in the process includes the identification of training needs, delivery of the required training and testing of the procedures and protocols discussed in training. There are several mechanisms that can be used to accomplish this last objective of testing the materials learned.

The first method and the one that is the focus of this chapter involves tabletop training. Tabletop training has both advantages and disadvantages over hands-on training. Advantages include the ability to use the principles of *modeling, or simulation*. These concepts allow trainers present the training scenario without requiring the presence of all of the resources. The benefit is that the agencies involved can apply certain training concepts, without needing to pay responders to participate during the drill. The financial saving, including the reduction of associated costs such as equipment, is a major advantage to this type of drill. The other advantage is that this type of drill can be limited to those in command and general staff positions under incident command protocols. This allows these responders to work together under semi-realistic conditions, which enables them to forge the relationships that will be tested during real-world applications.

The disadvantages are the opposite of the advantages in this case. Primarily, only the command and general staff positions need to be filled. Rank and file members of response organizations are not needed. As a result, these responders lose the opportunity to work together and develop a bond of familiarity at actual responses. Also, their ability to make adjustments to existing plans cannot be tested and recorded, since conditions are being simulated. These factors can be addressed by conducting hands-on drills, which will be discussed in the following chapter.

Tabletop exercises can also be used as a precursor for hands-on training, and can prepare those in command and in general staff positions for eventualities and circumstances that will be tested during the hands-on training evolutions. Both types of training have their place, and both will be explained. The first step though, is to conduct a tabletop drill.

Purposes and Goals

As with any type of training, or testing of an annex, there must be a purpose and a goal or a set of goals. These need to be developed early in the process because they will provide the framework for the testing. These goals should be terminal in nature and also include enabling objectives. The enabling objectives will be specific to the incident or event chosen. For example, a fire department conducts a drill involving live fire training. The terminal objective may be to provide live fire training in a live, yet controlled environment. The enabling objectives may include testing the Incident Command System, providing for the safety of all participants, and focusing on proper search techniques, proper suppression techniques, etc. These goals provide a general overall goal, and a series of specific goals to be tested during the drill. Therefore, goals for the drill need to be developed and documented, and the drill should be designed to test these goals. So, for the tabletop, the terminal objective may be to test the ability of Smithville responders to respond to and mitigate a chemical nerve agent at the Town Mall. The enabling objectives may include establishing the ICS, requesting and utilizing mutual aid responders, providing for the safety of responders, and establishing a command post. Whatever the goals, they must be reasonable, measurable and realistic. There are no set number of minimum goals, nor are there set maximum numbers. The goals need to be clearly established, the drill must work to achieve these goals, and all participants must be made aware of them.

Tabletop Scenario Planning Steps

When planning a tabletop drill, there are a number of steps in the process that need to be completed. The person or agency organizing the tabletop drill must decide a number of critical issues at the beginning of this process. The initial decision will entail choosing an incident or an event. These are two distinct happenings and each has different dynamics regarding pre-planning. An event is something which is planned in advance, such as fireworks displays, concerts, or parades. The beneficial aspect of events is that planners will know when and where the event will take place weeks, months, even years in advance. An incident, on the other hand is a happening that occurs without warning, and which may have a dire outcome. Incidents may include fires, natural disaster, and hazardous materials release. While agencies may have general plans for dealing with incidents, they cannot determine where each incident will occur. OEM, police and fire agencies can anticipate likely incidents at fixed facilities such as chemical facilities, nuclear facilities, hospitals, schools or banquet type facilities.

The concept of *pre-planning* involves certain agencies analyzing a hazard or threat and producing a plan of action designed to mitigate anticipated issues at any facility. For example, if a police department has a prison in its jurisdiction, the department would be wise to pre-plan for a number of potential incidents that may occur inside the prison. These may include fires, riots, escapes, and even natural disasters that require aid to be given to assist the corrections department. Anticipating, as well as planning appropriate responses to these incidents will make dealing with them much easier if the need arises.

Firefighting agencies are generally well versed in pre-planning for incidents. Their pre-plans usually include owner contact information for a facility, utility controls, fire and burglar alarm information, information regarding special hazards inside the structure, type of construction, occupancy of the facility, and hydrant locations as well as water flow capabilities. In some cases, apparatus positioning and preliminary tactics can be included in these plans.

Pre-plans are written and copies are maintained in a database and in response vehicles. These plans should be reviewed periodically to ensure accurateness and to maintain current information. Responders should also review these pre-plans, familiarize themselves with the response plans, and be prepared to act on them.

Incident versus Event

The tabletop planned for this chapter will involve a terrorist attack using chemical nerve agents and will be simulated to have occurred in a crowded mall. Planners at this point may opt to plan the incident themselves or they may seek guidance. Any number of people may be available to assist in planning for this drill. Choosing a person with experience in the field can provide many benefits. Experts can indicate dissemination parameters, expected purity levels, weather effects on the scenario, and expected casualty estimates etc. Choosing an expert will add realism to the drill and will allow for a realistic evaluation of the actions and decisions being made.

Drill Specifics

For the drill in this chapter, there will be a total of fifteen fatalities, and 600 other victims, with varying degrees of injuries and medical complaints. The attack will take place on a Sunday in the fall and will occur at 1:00 P.M. The weather will be clear with temperatures hovering near fifty-five degrees and wind gusts to fifteen miles per hour, from the west. The main entrance of the mall faces north, and the attack will be in the center of the mall, just inside the food court. At the time of the release, there will be several hundred shoppers stopping for lunch, and shoppers and mall employees will take all available seating in the food court. This will be similar to the incident previously discussed in this book, which will prevent me from having to explore multiple attack scenarios, and the possible responses to each.

We have now identified that we will drill an incident, which is criminal in nature and involves civilians with expected casualties. The town to be attacked in this drill will be called Smithtown. Smithtown is a small suburban community, approximately thirty-five square miles with a daytime population of 35,000 residents. Smithtown has three schools and a volunteer fire department with three engines, one truck, and a rescue truck. Smithtown's first aid squad is also all-volunteer and equipped with three BLS ambulances. ALS responders are under the control of the local hospital, located twelve miles outside the corporate limits of the town. The response time to the hospital via ground transport is eighteen minutes. The police department consists of forty-seven members, with patrol officers each working a twelve-hour shift with three days on, and three days off. Each shift has nine officers assigned to it as well as officers assigned as detectives, and supervisory and administrative personnel. Smithtown has its own Office of Emergency Management (OEM), staffed by three part-time volunteer workers. The town has its own water company, and department of public works including garbage removal crews. Utilities are provided by outside companies that provide natural gas, electricity, telephone and cable television to the town. The board of education has a fleet of twelve school buses. Smithtown is located in Essex County, which has its own OEM and other emergency resources including the hazardous materials responders for the county. Essex County covers 370 square miles, and has eighteen municipalities within its borders. Smithtown has filed its revised Emergency Operations Plan (EOP) within the last three months, and the county has approved it. The mall is the only mall located in the county; it is enclosed and contains a total of sixty independent stores and a food court all located over two floors. Now that the scenario has been chosen, the next step in the planning process will be to decide which other groups and agencies should be invited to participate.

Who Else Will Attend the Drill?

Given the incident, a number of agencies should be considered as training partners for this drill. We have identified the incident as a terrorist attack on a mall in which terrorists use chemical warfare agents, namely nerve agents. With this information in hand, we can begin to develop our list of response agencies. Since we will have issues with chemical agents and product identification, the county hazardous materials team needs to be invited along with the local fire, police, EMS, hospital staff, local and county OEM. This base will need to be expanded and, as the incident unfolds, state officials from OEM, hazardous materials response, and even public health will be invited. In addition to the local responders mentioned, all mutual and automatic aid response organizations will need to be invited. These resources will include fire, police, EMS and other responders. Since the incident will be reasonably large in nature, additional mutual aid towns will be invited, as will OEM personnel from all surrounding municipalities and possibly other counties.

Selection of Outside Observers

Once decisions are made regarding the participants, it would be wise to arrange for professional responders to serve as observers. These observers should be from agencies that are not participating and, in addition, from various jurisdictional and professional response backgrounds. Both state and federal level observers are good choices for these types of drills. Organizers should strive to choose observers with no direct or indirect ties to any of the participants. This will help ensure that the observers can remain unbiased, and provide thoughtful and honest feedback, without fear of hurting the feelings of friends of coworkers. The observers, who will play a critical role both during and after the drill, need to be able to speak freely.

You may also choose several moderators. The moderators will review the scenario, and work separately to develop changes to the incident. These changing parameters should not be discussed during the planning stages. During the drill, the moderators will have input and will inject information or provide changing circumstances. This is to test the abilities of those in charge regarding the timeliness and definitive nature with which the issues presented were resolved, or at least properly handled.

Choosing the moderators and observers must be done early in the process, and qualified people need to be put in those positions. The ability of the moderators to infuse real-world problems into the incident will be critical to the goal of testing certain responders, or annexes during the drill. Examples of details that can be used include responders being out sick that day, equipment failures, lack of volunteer staffing, sudden weather changes, secondary attack scenarios, and delayed responses from certain agencies. This type of information added to the drill without the prior knowledge of the participants will provide an even better analysis of their ability to work through the potential issues presented.

Incident Management Structure Development

The next step entails the development of an incident command structure. At this time, there are two basic command structures that can be established for this type of incident. The first structure involves a system a single incident commander (IC) for the incident. Under this structure, one person would ultimately be responsible for the entire incident and would make all related decisions regarding mitigation issues with the incident, including priorities and strategies. The sole IC would have the authority to establish the other command and general staff positions that he or she would require. Our drill is a very involved scenario and a single IC may be overwhelmed very early into the incident.

The second possible structure is that of a unified command (UC). Under this structure, a group of qualified people share the responsibilities of command. There are times where unified command may be legally required such as incidents which cross jurisdictional boundaries or other instances where a unified command structure is more logical. These incidents may involve more than one group or agency with functional responsibility, or when an incident will be expanding and too complex for a single IC. For example, a school bus accident involving a hazardous materials truck and the entrapment of children on a heavily traveled roadway may require a UC. There will be issues of traffic (law enforcement), entrapment (fire department), hazardous materials (hazmat), triage, treatment and transport (EMS), along with other issues including roadway, spill cleanup, notification of the parents, etc. Each of these groups will have certain fundamental and tactical concerns associated with this incident. Therefore, a unified command structure would be the best option, since all areas of concern by all responders can be addressed via UC. Under UC, several qualified people will act in concert to establish priorities and strategies to mitigate the incident. The UC will designate one spokesperson to be the *voice* of UC, and this person will speak on behalf of the group. The advantages of UC include the presence of trained people with special areas of expertise who can assist in developing strategies and priorities. Also, the other members of UC can serve as a sounding board, and the ability to have a greater operational picture is advantageous. The liabilities of UC include the premise that too many chefs may spoil the soup, and also that the members of the UC may be unable to come to a decision, which may hinder or slow the mitigation of the incident. All UC members must be able to *check their egos at the door* and agree to work together to resolve the incident. One possible stumbling block can be a situation where members of the UC are unable to come to a common ground. This scenario will delay critical decisions and can have serious repercussions. The group must be willing to see the big picture and agree to work together for the benefit of all involved.

Unified Command Chosen

For our tabletop, the planners will decide on a unified command structure for their incident management system. This incident is too large and complex for one single IC to manage alone. The next step in the process will be to construct the ICS, and determine who will be invited to be part of the UC. In this case, the following agencies should be represented as part of the UC. From Smithtown, we will include the following: fire, police, EMS, OEM, and the business administrator. From Essex County, we will include OEM and an officer from the hazardous materials team. This will be the basis of the unified command structure to be put in place for the initial response to the attack at the mall.

As the incident unfolds, there is always the possibility of expanding or contracting the command structure, including the makeup of the UC. This is not to say that this is the only way or that it is the best way to staff the UC. This is an example of one possible structure for the incident management option chosen for this scenario. Later in the incident, federal authorities as well as state authorities who respond will be integrated into the UC under various authorities.

Continuing the development of the IC structure, planners must determine the scope of the system to be used to manage this incident. As noted previously, there are seven major components of the ICS. Planners will decide if they will use all general staff and command staff positions during this incident. They will need to keep in mind that whatever sections and command positions are opened will need to be staffed. The incident that is the basis for this drill is fairly complex and fairly straightforward as well. Certainly, there will be a need for operations, planning and logistics. A decision must be made regarding the need for the finance/administration section for this incident. I will recommend that the finance section be opened, as there is a tremendous potential for the units within that section to be required. As for the command staff, I would also recommend that all three positions be filled. These include the safety officer, public information officer and the liaison officer. At this point, we know that we will need to staff all seven general and command staff positions. The next decision will be to decide who will lead each section.

Planning Section

The planning section chief (PSC) must be well versed in all aspects of the ICS. Additionally, the chief must be able to see this incident from all angles. The plan developed must take into account every aspect of this incident from responding to the location, tracking each resource throughout the incident, and lastly to demobilizing all resources. Several units under the planning section will need to be opened, and will include the resources unit, situation unit, documentation unit, intelligence unit and demobilization unit.

The planning section will develop the incident action plan (IAP) that will provide the basic response plan for the incident. This document has several forms attached, including a medical plan, communications plan, demobilization plan, and overall tactical plan for mitigating the incident. Each of the units under the PSC will need to be properly staffed. For the planning section alone, five supervisory positions will be needed, not including backup or support positions. Supervisory personnel in the ICS are referred to as *overhead*. Each unit will require a unit leader, and the planning chief will need a deputy. Therefore, a need for seven overhead positions is required for the planning section alone.

Logistics Section

The next Section that needs to be considered is the Logistics section. Again, there will be a logistics section chief and a deputy. The units under logistics that we will fill for this drill include the food unit, medical unit, communications unit and ground support unit. Therefore, there is a total overhead need under logistics of six personnel.

The planners will need to determine how best to staff these overhead positions for the drill. In county OEM, there are usually bureau chiefs that serve as liaisons for each bureau under OEM. For example, there will be police, fire and EMS bureau chiefs. These chiefs will serve as liaisons to the logistics section chief (LSC), and provide real-time personnel and equipment availability and response times.

Finance/Administration Section

The finance/administration section will require a chief and a deputy. For the purposes of this drill, finance will open a cost unit, and compensation claims unit. The other units located under finance will not be needed for the purposes of this drill. Accordingly, the overhead need for the finance section will be limited to four.

Operations Section

The operations section will have the largest number of responders and resources. There will be an obvious need for a deputy chief for operations. However, at this point in the planning process, a number of decisions will need to be made. Operations will have multiple *branches* operating under its control. For this drill, they will include law enforcement, fire, hazardous materials, EMS, and staging. Each branch will have a branch director, and beneath the director will be various groups or divisions. Under the ICS, groups are bodies of resources with a certain *functional* responsibility. For example, police would be assigned to the law enforcement group, medical personnel to the EMS Group, etc. *Divisions* are where resources are divided into areas of *geographic* responsibility, such as the west division, or the north division. In a division, you may have different functional groups. For example the law enforcement group may have a traffic control component, and this component may be assigned to the east division to perform the traffic control. Therefore, under operations, there will be five branches. Each of these branches will require the further breakdown of resources to work, called groups and divisions. Each of these groups and divisions requires overhead.

The leader of a group or a division is called a *supervisor*. If, under the law enforcement branch, we need a security group, a traffic control group, a perimeter security group, a patrol group and a relief group, that branch will have a total overhead need of six personnel. These would include the branch director, and the five group supervisors. If under the EMS branch, there were a triage group, a treatment group, a transport group, rehabilitation group, and an aid station group, the overall overhead need for the EMS branch would also be six positions. These positions will include the branch director and the five group supervisors. Under the fire branch, for our drill, we will require a search and rescue group, water supply group, air bottle refill group, ventilation group and rapid intervention team group. The fire branch will also require a total of six overhead positions, including the branch director and the five group supervisors. The hazardous materials branch will have a decontamination group, air monitoring group, and product identification group. Therefore, their total overhead need will be four, including a branch director, and three group supervisors. Lastly, the staging area will require one overhead position per staging area. The overhead for the staging area is called the staging area manager, and this person should have an assistant, bringing the total overhead need for each staging area opened to a total of two positions.

Command Staff Positions

The command staff positions of safety officer, liaison officer and public information officer will all be utilized for this drill. The overhead for each position will include the officer, and an assistant, bringing the total overhead for the command staff to a total of six positions. For the entire drill there is a need for a total of forty-nine overhead positions, assuming that we do not have any need to expand the size of our existing incident management structure. Any additional groups, divisions or units will require additional overhead be requested, in order to maintain operational efficiency at the drill. Even filling only the overhead positions for this scenario, it is very obvious that the mitigation and rescue operations will require hundreds or thousands of responders. This is why a tabletop scenario can save money for those involved. The planners can simulate the bulk of those who would be doing the work at a real incident.

There is a need for forty-nine overhead positions in our incident management system for the purposes of this drill. But why is there such a need? The concept is called *span of control*. This principle holds that supervisors can only safely and directly supervise a certain number of subordinates. Exceeding this number places the resources being supervised in potential danger, because of the lack of close, direct supervision. Under the incident command system, the range for span of control is between three and seven subordinates. Less than three elements do not require an independent supervisor, and supervising more than seven is deemed unsafe.

Another concept at work in this system is that of *unity of command*. This principle holds that each member working on the incident answers to only one direct supervisor. For example, consider the following scenario. The operations chief is responsible for all resources and personnel working under that section. In the case of our example, there are five branches. Each branch director answers directly to the operations section chief (OSC). Under the law enforcement branch, we have established five groups. Each of those groups has a supervisor. The group supervisor reports directly to the branch director, who reports directly to the operations chief. In the traffic group, there may be four divisions, each with a supervisor. Each of the four division supervisors will report to the traffic group supervisor, who reports to the branch director, who reports to the operations section chief. We can keep this example going, to show hundreds of personnel in the law enforcement branch. However, each single resource will report only to one supervisor, who will report to only one supervisor, all the way back to the section chief.

Remember, the section chief only has direct contact with the five branch directors, regardless of the number of resources working under each branch. Similarly, the only person the operations chief answers to is the IC, or the UC. The IC/UC only has direct contact with the general staff, and the command staff. The total maximum number of overhead that the IC/UC will have direct contact with is seven. This number represents the maximum number of reporting elements under the span of control doctrine. Regardless of the size of your incident command structure, every person there has only one person to whom they are obligated to report. Strict adherence to these two principles will help to maintain the safety and accountability of all responders working at your incident. Planners need to have the correct numbers of overhead in relation to the overall number of subordinate positions in need of supervision.

Applying Modeling and Simulation

As has already been noted, one advantage of a tabletop exercise is that planners may use modeling or simulation to reduce the number of participants and to save money. At this point, those planning the tabletop drill must determine which overhead positions will be filled, and which will not. This is a critical decision making process, since the drill moderators have the ability to insert scenario changes during the drill. Planners need to determine the exact number of persons that will make up the ICS for the drill. It would be wise to be sure that the players involved in the drill have real-world authority and responsibility. The purpose of the tabletop is to test an annex or a scenario without the overall manpower requirement that will be needed for an actual event. Planners need to envision the scenario prior to determining which overhead positions are to be simulated. Overhead should be filled to at least the branch director level, and certainly the various group supervisors can be simulated. For this drill, we can simulate the hazardous materials groups, and fire groups as well. Most of the law enforcement groups and staging can be simulated. The EMS groups can also be simulated. Therefore, if there is a need to reduce the ICS structure due to manpower requirements for the drill, the planners can shave off twenty-five or more of the forty-nine overhead positions. Planners should make every effort to staff as many of the overhead positions as is feasible, since there will be no actual tactical operations. Involving as many overhead responders as possible will provide greater numbers of responders to get involved in the drill. Since there is a possibility that many of the overhead positions may be filled by responders with no real ICS experience, it would be advisable to involve as many of them as possible. This involvement will allow them to see how the ICS structure and unified command will work It will also permit them to provide their own input, as well as to ask questions regarding their individual responsibilities at the incident based upon their assignment. Involving all of the agencies and jurisdictions that would respond to an actual event will generally provide enough access to personnel to staff all of the required overhead positions.

During training evolutions that involve many agencies and responder groups, it is important to allow the players in the drill to work within their area of specialization. If tabletop scenarios are a routine occurrence in your jurisdiction, roles may be exchanged for purposes of training. For the initial and first few scenarios, the best course of action would be to allow the police to man law enforcement overhead positions, fire to staff the fire overhead positions, and so on. In this manner, the responders can observe the functioning of the ICS structure while focusing on their areas of expertise. For subsequent drills, overhead staff may be rotated into other areas to broaden their background and experience. Therefore at later drills, the law enforcement responders for example may be rotated into the planning section, to assist in authoring the IAP. One purpose of training is certainly to expose the trainees to different information, and to provide them with an opportunity to expand their operational capabilities. This goal is best served by ensuring that trainees are extremely familiar with their day-to-day responsibilities prior to learning a new skill set. Accordingly, for this drill, members of the fire service will staff the fire branch, and members for that service, etc., will staff the hazardous materials branch. During the drill, these overhead positions being so staffed will allow for a prompt response to any additional changes to the scenario, which may be introduced by the drill moderators.

Staging Areas

There is another aspect of the ICS structure that needs to be reviewed and tested during the drill. Under the operations section are a number of possible staging areas. During the event or incident, staging areas should never be empty because they have certain responsibilities during the incident. These areas allow for the incident managers to have additional trained and staffed resources immediately available for use during the incident. Without staging areas, there would be a delay in ordering, dispatching, and receiving critically needed resources. Staging areas provide ready access to resources which are anticipated to be needed during the incident.

Under the principles of the ICS, there may be any number of staging areas for the incident, but each must meet certain criteria. The staging areas need to be given separate names; must be secure locations preferably with a power supply; have rest room facilities; and provide the facilities necessary to refuel the vehicles and apparatus staged there. Staging areas need to be in close proximity to the incident location and response times need to be less than five minutes when exiting the staging area. Rest room facilities may be brought to the staging areas in the form of portable toilets, such as are seen in parking areas at stadiums and arenas. Another procedural aspect that occurs in staging areas is the forming of strike teams and task forces.

To review, a *strike team* is a group of the same kind and type of resources. A law enforcement strike team may consist of three to seven police cars, is required to have its own transportation, common communications, and a leader. Strike teams can be composed of any of the single following resources: fire engines, rescue vehicles, construction equipment, ambulances, etc., so long as they are the same kind and type. These teams have specific duties and responsibilities. A large fight would require a strike team of law enforcement responders to be dispatched.

Conversely, a *task force* is composed of resources that are not the same. For example, inclusion of a fire engine, fire truck, fire rescue vehicle, BLS ambulance, two police cars and a hazmat vehicle would make up a task force. The advantage of a task force is that it can be deployed, and may be self-sufficient for a number of tactical duties. At a house fire, the task force has all that would be required for a first alarm fire; additional resources would not be needed. The dispatching of a strike team would not be appropriate at this type of incident, since multiple responders across several response disciplines would be required for proper resolution of the incident. During our drill, overhead staff may consider whether or not to form strike teams or task forces for prompt deployment from the resources at the staging areas.

Organizing Resources into Strike Teams and Task Forces

For the tabletop we have decided on, the planners need to decide if they want to construct strike teams or task forces. Given the immediate need for certain resources such as BLS responders and ambulances, resources may not be available to form strike teams. Initially, most of these assets will be sent to the scene or to staging for rapid deployment once their need is established. For example, with a great number of victims, comes a great need for BLS Units. If the treatment group has completed field treatment of a patient, the BLS Unit will be required to transport that victim to a hospital or other facility for definitive medical care. If BLS Units are staged, they can be deployed as requested to a specific location for the transport. Another option would be to have all of the BLS Units report directly to the scene. Ultimately, the decision as whether to stage the BLS Units in a staging area or to have them report to the scene would be made by the branch director, group supervisor or operations chief. Regardless, planners will need to determine if the resources will report to staging, or any other location. Also, the decisions must be made regarding the establishing of strike teams or task forces.

Law Enforcement Branch

Another area to be discussed prior to the actual drill involves the law enforcement branch, which will specifically address security controls, including perimeter access and control. The planners will need to consider the development of control zones— hot, warm, and cold—and security controls for each. Generally, the hazmat teams determine the control zones, but security concerns also need to be considered. There must be adequate security in place at the command post. This security control needs to be addressed under the operations section chief, and eventually through the law enforcement branch. Certain emergency operations plans (EOP) address the security needs of fixed and mobile command posts and determine in advance which agency will provide this security.

If the security is not addressed in the EOP, the command structure must address where the security will come from, and what the protocols will be. Law enforcement officers will need to be made aware of the security requirements and concerns at the command post, such as the type of identification required for entry or possibly a list of those approved for entry. This information needs to be discussed and developed early in the incident to allow the proper implementation by security personnel.

The law enforcement branch will also be charged with security control at the inner and outer perimeters, or the warm and cold zones. Again, during a tabletop scenario, these positions may be simulated in order to reduce manpower requirements and to save money. The development of security controls for the scenario are critical, and during actual incidents will need to be quickly put in place to secure the scene of the incident.

Pre-scenario Review

At this point in the process, we have chosen an incident, developed an incident management system and structure down to the group or division level. We have identified the overhead or supervisory staff needs and we have also decided which of these positions may be simulated. We have put a unified command team in place, and we are ready for the next step, which will involve reviewing the scenario to be used with all of the participants.

Participants should be given a packet of information for review and comment prior to the actual drill. Information in this packet will include a thorough explanation of the date, time, weather, occupancy, and location of the target location. The type of incident should be revealed, but certain information needs to be withheld to properly test the responders. There are usually several meetings wherein the finalized incident has been reviewed and revised until all agree it is ready for testing.

The information withheld will include product identification, amount, and purity as well as dissemination parameters. The packet should also contain information concerning the ICS organization and the positions to be filled, as well as those being simulated. A complete listing of filled positions needs to be placed on an organizational chart, which should also include the names of all responders with positions that are being filled. A timeline also needs to be developed and careful consideration should be paid to it. Realistic time frames must be included into the drill in order to keep the scenario moving forward as it will during an actual incident.

Sample Timeline for Drill

The time line is critical and must reflect all steps that will be taken at an actual incident. Below, I provide a brief sample timeline, but it will not contain information that will be added to the incident by the moderators. The roles and duties of the moderators will be discussed and reviewed following the sample timeline.

Sample Initial Timeline for Smithtown Mall Tabletop Drill:

1300: Mall security gets reports of odors in the food court and ill customers.

1304: Mall security guards arrive on scene and confirm odors and illnesses. Security staff reports a half dozen victims choking and coughing.

1305: Mall security contacts 9-1-1 center, requests police, fire and first aid.

1307: Additional security personnel respond to the food court. They report additional victims, including the first arriving security personnel. Further, they report mass panic and shoppers fleeing the area.

1310: Police units arrive on the scene and enter the mall. They report more victims, and encounter victims trying to exit the mall, but not making it all the way out to safety. They retreat.

1318: Fire department on scene, Fire chief establishes Smithtown Mall Command. Hazmat and a second alarm are requested. Fire crews don SCBA and standby, ready enter the mall if cleared. No rescue operations are underway.

1319: First BLS Unit arrives, and begins treatment of victims who self-evacuated.

1322: Command requests a full third alarm of fire resources and fifteen more BLS Units.

1323: Additional police units are on the scene. Shoppers from all areas are fleeing the mall en masse. Second BLS Unit arrives; police try to establish security and traffic control.

1325: Additional fire units and police units arrive.

1330: Essex County Hazmat arrives, and meets with IC. UC is quickly established.

1334: Media personnel are beginning to arrive, and are asking for information.

In reviewing the sample timeline above, you will note that it contains only information, with very little in the way of decisions that were made. The purpose of the drill will be to provide a basic framework of information to be provided to the incident managers. These responders will then make decisions that will be noted in the body of the timeline. For example, note that the firefighters were banned from entering the building. Why was this done? Simply put, their structural fire fighting gear is not designed for prolonged exposure to chemicals. They cannot enter until they receive other Class A Personal Protective Equipment (PPE) ensembles, or the Hazmat team identifies the agent and it is determined that the structural ensemble with SCBA is safe for the crew to enter. Hazardous materials team members must be sent in wearing appropriate PPE, in order to identify the product released, to monitor the air, or to continue rescue operations. The UC must decide what tasks will be assigned to each resource. These decisions are the crux of the tabletop drill.

Introduction of Additional Facts into the Drill

The moderators will have been at the meetings during which the scenario was reviewed. During the drill, the moderators will meet, and using the timeline, will gradually release additional information and facts to the scenario. The information will be added as deemed appropriate by the moderators and real-time responses can be tested. Some information will be handled via simulation of resources, and some will require action to be taken by either UC or branch directors or group supervisors. The information must be realistic and specific to the scenario. Information to be added may include product identification, additional symptoms exhibited by victims, secondary incident, non-availability of resources requested, etc. Caution needs to avoid overdoing the addition of aggravating factors to the incident. The purpose of the additional issues is to test responders' and managers' abilities to recognize the new information, and act accordingly to mitigate it.

The next step in the finalization of the drill will be to arrange for people to serve as observers. These observers need not be present during planning sessions and, in fact, may be better served by not attending, but they must be on hand for the entire duration of the drill. Failure to attend the planning meetings will prevent their coming into the drill with predetermined opinions regarding those involved. They will also be unaware of the perceived strengths and weaknesses as determined by the drill participants. The observers will need to have access to the drill information as distributed to the participants, but they do not need to be given the additional information or facts that will be introduced by the moderators.

Planners need to choose the right people to serve as observers. There are many qualified people who may be asked to serve as observers. They should be from the response community, and may represent local, county, state and federal agencies. The observers should be seen but not heard during the exercise, but should be taking copious notes, and be prepared to discuss what they see. It is often a good idea to choose observers with no ties to any participating organization, or even any individual who is participating in the exercise. If these parameters are followed, the observers should be in a position to offer completely unbiased opinions, without the fear that they will be passing judgment on a friend or a colleague. The ability to present opinions to the participants without feeling that they are hurting friends or co-workers is of paramount importance. The input of the observers will afford the participants the professional and honest evaluations of the performance of the drill participants. This honest feedback will be the basis for the review, corrective actions that need to be addressed, as well as the after action adjustments to existing policies and procedures. Observers must be chosen who can provide honest and accurate opinions of the exercise.

Conducting the Drill

After all of the planning meetings have been conducted, a date is chosen, and observers have been selected, all that remains is to conduct the exercise. All players in the drill should be reminded that they must report where they are directed and will need to be there on time. Once all of the players are accounted for, the drill can begin. The planners should begin by reviewing the rules of the exercise, and asking participants if there are any lingering or new questions. Once all of the preliminary information has been delivered, the drill can begin in earnest. Players must pay close attention to the timeline, since this is the document which will serve as the script for the exercise. At certain points during the exercise, moderators can infuse their information into the exercise, and incident managers will be allowed to respond to the changing course of events. Moderators can also ask questions of the incident managers, to test their ability to recognize issues that may occur during a real incident. Observers will take in all of the actions taken, along with the decisions and reactions to information added by the moderators. The goal at this point would be to see how the Unified Command structure is able to work together, and how they adapt to changes in the drill that were not anticipated. The exercise will probably take several hours to complete and does not have to be done in real time. For example, once the initial call is made, participants can "speed up" the process to eliminate waiting several minutes for responders to get on-scene. The objective will be to see how each participant responds to the duties they are asked to perform, and to determine the effectiveness and decisiveness of the UC. Those in overhead positions will have decisions that need to be made, and once the decisions are made, the observers will note for comment the speed with which they were made later. Contingency plans for the exercise should be tested as well, so that they too may be independently evaluated by an outside agency. Once the moderators have exhausted all of their additional information, and the simulation of the drill reaches its end, the exercise can be terminated.

Post-Drill Activities

At the conclusion of the drill, all participants will be given an opportunity to review their notes, and ask any questions. Moderators and observers will begin to organize their notes and observations. At this point, the planners of the drill will need to schedule a date, time and place for the formal process of reviewing the drill. There should be adequate time to allow for all involved to reflect on what occurred at the drill, and to prepare questions or raise concerns based upon the actions taken during the drill. The review, once scheduled will take several hours, and should be very thorough, but conducted informally. The review needs to reinforce the positive aspects as well as address deficiencies. Caution needs to be taken to remind participants that the review is not being conducted solely to shed bad light on any person

Figure 12.1 **A parking lot can be an excellent site for an experienced terrorist. A primary attack can be planned and executed, and the parking area will serve as a secondary attack. Once initial attack survivors flee to the "safety" of their vehicles, a second attack is planned. In this case, one or more well-placed car or other bombs will multiply the number of victims. Responders need to remain vigilant of potential secondary or tertiary devices in large, unsecured parking areas. (Image copyright Wizdata, Inc. 2009, under license from Shutterstock.com.)**

or group of people, but rather to allow for an honest assessment of the performance of all involved. Often, some people whose performances are being reviewed seem to take the evaluation personally, and this may be evident in this type of review. Leaders of the review need to remind them that the information gleaned from the drill will allow for correcting issues that are raised, and will not reflect poorly on anyone. I once participated in a hands-on drill, and during the demobilization process, drill organizers asked me for a few quick observations. We too simulated a CWA attack on a mall, and the majority of the responders were volunteers. One observation I made was that we needed more ambulances on-scene for the scope of the incident. At that point, the woman in charge of the local first aid squad became very upset, indicating that she could only provide the number of ambulances that she had. Obviously, it is unrealistic to think that a municipality will purchase fifty ambulances *in case* calamity strikes. The point was that we need to order more from our mutual aid partners, but the woman took the fact that she has only six on hand to be some kind of condemnation of her leadership. This type of reaction may discourage moderators and observers from full disclosure for fear that participants will be upset. This will render the drill useless. Open, fair and honest reporting by observers and moderators is critical to the success of the drill.

Drill Review Sessions

During after-incident reviews that I have attended, the range of formality was very diverse. Some included PowerPoint presentations, some included the distribution of handouts and some were basically verbal reviews, where comments were solicited, and discussions ensued. After a drill of this scope, a formal review would be preferred, but conducted informally. Using either a handout or the PowerPoint method would probably be best, since participants can follow along, and the review would be organized. The method of review is not as important as the content of the review. The timeline should serve as the basis for the review, and each decision that was made, or not made needs to be commented on. Reviews should emphasize the good and the bad. Glossing over the good or innovative decisions is bad, and credit needs to be given when the decisions and actions were appropriate and correct. Each decision need not be labored over, simply commented on. Moderators and observers will have specific decisions that they will focus on, and discussions may ensue. Moderators specifically will be interested in the decisions that were made to information that they added to the drill. They will, therefore, be in the best position to comment on the decisions that were made by the participants. Accordingly, the input they offer regarding this information that was added to the scenario should be given its due attention and consideration. During the review, so long as the process is orderly, all participants should be allowed to offer their input. The decision to allow participants to get into debates over decisions that were made is up to the person or people who are in charge of the review. If every decision that was made is subject to review and debate, the process will break down, and the benefits of the review will not be fully realized. If there are certain critical decisions that were made, and in need of review and discussion, that process would certainly be in order. Those running the review must not allow the review to be a personal review, where participants feel compelled to explain every decision that was made.

The review process must be kept on track, and moving forward at all times. Attention must be paid to preventing small *breakout groups*, or groups who will debate certain points while the review is still in progress. If these side conversations are allowed, the process will erode, and once again, the full benefit of the review will be lost. Those who are serving as proctors of the review will need to be tolerant and patient, since there may be participants who are desirous of seeking detailed explanation of certain decisions or actions that were undertaken. This part of the process gets very tedious, but skillful proctors will allow all questions and inquiries to be addressed, while keeping the process moving forward. Once the review is completed, the next meeting can be planned, wherein the recommendations of the reviewing members can be acted upon, and policy changes made.

Time needs to be allowed between the formal review process and the next meeting where corrective actions are to be addressed. Unless the drill was flawlessly executed, there will be a recognized need to alter, amend or create policies regarding certain aspects that were uncovered during the drill. Once the date for this meeting is established, senior members of the agency and those with policy-making authority need to be assembled. The moderator of this process needs to be very well organized for this meeting. As a result of the formal review, policy makers should have made a list of issues in the existing policies that are deficient. Further, any new issues brought up during the review that are not subject to a controlling standard operating guideline need to be addressed as well. Those with the ability to propose, author, or approve policy and procedural changes need to be at this corrective action meeting. If they have questions or concerns regarding the proposed policy amendments, time needs to be provided to explain why the revisions are being proposed. Examples of policy changes may include issuance of additional or new Personal Protective Equipment (PPE), issuance of respiratory protection, establishing an accountability system for operations during certain responses, limiting exposure time during responses, etc. Many times, these policy adjustments or the creation of new protocols will require training for the responders. This needs to be carefully considered when making these requested changes. If changes will require training, there must be a mechanism in place to provide for this new training. In other cases, the revisions to policies and procedures may only require language changes, or other administrative changes or updates. Given these changes, additional training will not be required of anyone.

After-Drill Adjustments

In cases where an annex in an Emergency Operations Plan (EOP) needs revision, the revisions must be made, signed off by the appropriate parties, and forwarded to the county for final approval. These annex adjustments also may require training, or possibly simply will be modified with new contact numbers, or contact responders. Either way, the annexes that were affected by the drill need to be reviewed, and accuracy and currency in the existing plan need to be confirmed. There may have been several annexes that were impacted during the tabletop drill. OEM personnel need to be certain that all annexes that were used or tested are subject to this review. Once the revisions are approved, all OEM personnel need to be made aware of the revisions. OEM and response agencies should review EOP and response policies on a regular basis to ensure currency and applicability.

Summary

In summary, the planning of a tabletop exercise has certain distinct advantages, and also has limitations. The planning of a tabletop exercise is a prolonged process, requiring many participants and planners willing to take the time to follow all of the required steps in the process. Failure to follow all of the steps may prevent a complete and fair tabletop scenario, that would minimize the importance of the policies and annexes that were being tested.

One of the first decisions will be to determine if the scenario will involve an incident or an event. Once that decision is made, a specific scenario must be chosen. At this point, local experts may be consulted to assure that the scenario chosen is properly designed, and presented. This expert can assist in determining a timeline, and giving realistic developments, casualty estimates, responder needs and requirements, etc. Choosing to involve an expert is a decision that must be given careful consideration. The expert can ensure realism and can also assist moderators in determining information to be added to the scenario. This type of outside assistance can make the entire scenario and follow-up procedures much more realistic and, therefore, much more beneficial to all of the participants. The expert, or you, should you opt to act alone, will then choose the specifics of the scenario. These should include day of the week, date, and time, nature of the incident, casualty estimates, resource requirements, and other miscellaneous factors that may be included in the drill design. Planners will also determine which, if any annexes from the local or county EOP will be tested at this point.

Those in charge of planning the incident also need to decide which other agencies will be invited to serve as part of the drill. Depending on the scenario chosen and the scope of the incident, the invited participants may be limited to a few, or if the scenario involves a larger, more complex incident, a larger pool of participants will be required. Planners need to determine which response agencies will be involved in the mitigation efforts and which of their mutual aid or automatic aid partners may be interested in attending. Additionally, since they will be involved in an actual incident, responders from both county and state level agencies should also be asked to participate. In certain cases, there will be swift response to the scene from certain federal agencies and accordingly they should also be invited to serve as players in the drill. The state and federal assets can be introduced later in the drill, to provide a level of realism, since they will most likely not be on-scene immediately.

The invited players will expect that they have clearly defined roles in the scenario and planners may act to meet these expectations. In the scenario we developed for this chapter, the need for overhead personnel, or supervisors, was set at forty-nine. This is a large number of overhead personnel and sheds some light as to the total number of responders that would be needed to mitigate the crisis, should it be real. The number of overhead requirements may be too great for a first time tabletop drill. This need may be overcome by simulating some of these overhead positions, thereby allowing for a tighter group of decision makers for the drill. Simulation involves *pretending* that the position or responder is actually being filled, without actually staffing that position. This also allows for a more cost effective drill, since actual responders will not be required to be in attendance. Given this fact, there will be no need to backfill the position of the drill participant, which will reduce overtime expenditures for the drill. This ability to model or simulate responders on the day of the drill makes a tabletop scenario very desirable to administrators, who can drill without breaking the budgetary bank.

After the invited guests have signed on and the scenario is selected, assignments for all of the players need to be determined. Planners must establish a working incident command system (ICS) structure for the drill. Many positions need to be filled, at levels from section chiefs all the way down to group supervisors and leaders. Those in charge also need to choose whether the single or unified command structure will be used. Most likely, unified command will be chosen, and those invited to be part of unified command need to be selected. The balance of the structure needs to be developed and key positions need to be staffed. At this point, keeping responders in a capacity closely tied to their emergency responsibilities is desirable. Allowing them to maintain a certain comfort level by being assigned to a position that they are familiar with will make them more comfortable and help keep the flow of the drill moving. At later drills, the players may be rotated to areas where they have no expertise in order to allow them to focus on the scenario as a whole rather than their own little portion of the drill. This allows for better-rounded responders, which will be extremely beneficial at a real incident. All of the section chiefs need to be assigned, and they need to be put in positions where they can be of great use. Many of the units under each section may be used for the scenario, but these positions may be subject to simulation. Planners must ensure that all of the truly necessary and sensitive ICS positions are staffed and staffed in a manner consistent with what will take place during an actual incident. This part of the process will require tremendous thought and time, but will be critical at the end of the drill when the process is reviewed by all of the participants. Placing the players in positions consistent with their day-to-day responsibilities will allow for a true evaluation of their performance.

Another decision involves the use and placement of staging areas. Planners will need to decide if the resources required to mitigate this scenario will all be immediately used, or if they will allow some to be housed in staging areas. If the resources are staged, the participants will need to be tested on their knowledge of these ICS facilities. Each will need a distinct name and the drill can also test choosing the correct locations. Other aspects of staging areas, such as restroom facilities, food and water, security, and size of the area are subject to evaluation by the observers.

Observers also need to be contacted and they must be willing to participate in the drill in that capacity. Their purpose will be to be seen, but not heard. These people will evaluate the drill from all angles, and provide honest, frank evaluations of the participants, heir actions, and their decisions. It is best to select observers from local, county, state, and even federal levels and to choose participants who have no knowledge of the other participants. Also, it is best to not provide any policies or procedures that are in place, so they can give a full, unbiased grade to the participants. Knowing other participants and their protocols may prevent a fair and complete evaluation by the observers. Their ability to be completely honest without fear of reprisal or of hurting the feelings of anyone they know will be most beneficial. Holding back information or comments for any reason will take away from the overall effectiveness of the drill.

Finally, once the tabletop has ended and the review is complete, the agencies involved must make final decisions regarding policies, procedures, and updates. The exercise will most likely expose flaws in the agencies' policies and procedures. These flaws need to be accounted for in the after-action adjustment. In some cases, an entire annex from an EOP will need to be rewritten or substantially modified. Generally, these revisions and updates may be what prompted the exercise in the first place. If an agency identifies a response issue, a drill may be the best way to test the policy. Having competent, qualified observers and moderators will enable a fair and independent testing and evaluation of the procedures and policies currently in use by an agency. Planners must be willing to accept the assessment of the observers and moderators after the drill and must also be willing to address the underlying issues observed by the reviewing professionals. Problems will therefore be identified, and hopefully, corrected with due diligence.

As noted, the tabletop scenario will allow for training with reduced manpower. The timeline simulation for the incident used was basic and ended very early in the incident. During the planning and execution of your drill, the timeline will need to be considerably longer and much more comprehensive. Remember, the purpose of the tabletop can be to prepare for a future hands-on drill where less simulation and modeling will be allowed. These preliminary drills lay the groundwork for future drills and are an excellent way to get agency administrators, and supervisors ready for a large-scale drill. Tabletop scenarios are a step in the training process, and offer many benefits.

Chapter 13

Planning and Conducting a Hands-On Drill

In Chapter 12, the concept of the tabletop drill was explained and reviewed. It was noted that these types of exercises have advantages and disadvantages to them. One disadvantage is that the rank and file positions as well as several overhead positions are simulated rather than actually filled. While this will certainly save the drill planners and organizers money, those who will be working to mitigate the crisis will not receive the benefit of the training. Additionally, only the overhead position holders will reap any training benefit. Should an agency desire to have its responders tested in a specific area or in response to a certain scenario, a formal hands-on drill must be scheduled and executed. The process will be similar for overhead position holders, but will also involve the rank and file responders who will perform certain duties and functions. There are still many areas and functions that can be simulated during hands on training, but more responders will be subject to actual hands-on activities. As a result, their performances can be evaluated and rated including as well as the ability to react to fluid, ever-changing scenarios. The process, benefits and drawbacks to this type of training event will be discussed and debated in the balance of this chapter.

Modeling/Simulation for the Drill

Tabletop exercises rely very heavily on modeling or simulation and are generally designed to test the supervisors. Hand-on drills share some of these same traits, but also involve a great number of rank and file personnel. As with tabletop scenarios, the practical or hands-on training needs to be planned well in advance of the formal training. Since this drill requires additional personnel, the cost factor to execute the drill will be significantly higher. Planners and administrators must be aware of this prior to the start of the drill and expected costs, or cost analysis projections need to be developed and approved. These projections will be made after the very initial stages of planning are completed. The first decision will be to decide whether an incident or an event response will be contemplated. Again, for our purposes, an incident will be chosen, but there will be an added twist. The incident to be drilled will take place during a scheduled event. This will test the ability of responders to address contingency plans, to think on their feet, and to recognize and react to certain indicators. After the incident is chosen, as with the tabletop scenario, we must determine which response partners of the drilling agency will be invited to assist in the mitigation. Mutual aid responders, automatic aid responders, county, state, and federal responders should be considered as well. Moderators and observers, as explained in Chapter 12, will need to be selected, using the previously discussed criteria. Incident command structures need to be in place, and a date, time, and location of the drill need to be chosen. The drill can then be conducted. After the termination, a review date needs to be chosen and recommendations need to be considered and activated. Once those steps are completed, policies and procedures, along with annexes need to be amended and updated as determined by the reviewing policy-making authorities.

Choosing An Incident or Event

Now, we must decide on the incident for the drill. As indicated, this exercise will result as an incident that occurs during a scheduled event. The event will be a Fourth of July celebration and fireworks display being held in the Township of Salem. This is an annual event, and the planners have already developed a template for the event, and that template has been successfully used for the past nine years without major incident. The annual celebration will be in the Woodline Park, a municipal park located in Salem Township. The park is a 140-acre park with restrooms, soccer and softball fields, intermittent lighting, and picnic groves. There are five main parking areas, which can accommodate a total of 120 cars each and three smaller lots with a capacity of 30 cars each.

For the celebration, planners expect 5,000 spectators and partygoers for the day-long event. Planners expect all available parking spaces will be used and satellite parking has been established at the Salem High School. Buses will transport the overflow attendees to the park in fifteen-minute intervals beginning at 12:30 P.M. They will transport these same people back to the high school until 11:30 P.M. Other resources contemplated for this event include thirty police officers. These will include bicycle patrol officers, traffic control officers, plain-clothes officers, and roving patrol officers. A four-unit drunken driving squad will be on hand as well. Two lieutenants and five sergeants are also assigned to the field. The EMS resources include a total of eighteen EMTs. These responders will staff three aid stations;

will include three bicycle units with trauma kits; and have three BLS Ambulances stationed throughout the park. One ALS team will be assigned to the event as well. Included in the EMS responders will be a captain and three lieutenants from the local first aid squad. The local fire department will have two brush trucks and two fire engines assigned and will be located on opposite ends of the park. The brush trucks will have an officer and two firefighters and the engines will consist of an officer and three firefighters. Local OEM staff will be present in the command post, which will be located inside the maintenance building in the center of the park. The fire marshal will have the permits issued for the fireworks display, and his office will ensure that the fireworks are deployed under existing fire code provisions. Public works employees will prepare the park in advance of the celebration, and will add Porta-Johns® and additional trash receptacles throughout the park.

The weather for the day will be clear, humidity near seventy percent, temperature hovering near ninety degrees, with a southerly wind of ten miles per hour. The celebration will begin at 2:00 P.M. and end at approximately 10:00 P.M. All responders will be assigned to work a twelve-hour shift, beginning at 11:45 A.M., ending at 11:45 P.M. The incident will be an explosive attack, which will begin at 9:30 P.M. during the fireworks display. The initial detonations will come in the form of pipe bombs, which were placed in several garbage cans, as well as others placed in backpacks and left unattended. After thirty minutes, two car bombs will detonate in the large parking lots, one at each end of the park.

The event/incident scenario is one that should always be contemplated by both local planners and emergency responders. These types of public celebrations will always attract the attention of terrorists, whatever their social agenda. The 1996 Atlanta Olympic bombing illustrated the attractiveness of these events for terrorists. The publicity generated, the large mostly unprotected groups of civilians, and significant historical dates may be factors that lure terrorist groups to these events. Further, the relaxed atmosphere of these types of celebrations may result in lower than usual expectations of attack by terrorists.

Goals and Objectives of the Drill

Once decisions are made involving the scenario, planners need to meet and discuss the goals and objectives of this drill. These may include the ability to test the decision-making abilities of UC, the ability of street responders to react to a crisis, and the ability of EMS to triage a mass casualty incident (MCI), etc. Whatever the goals, the drill must be designed to set standards for assessing the responders responsible for the evaluation of the goals. If a goal is to set up a field hospital but the hospital or military are not involved, this goal may not be met. There must be a level of consistency between the goals and the parameters of the drill. Any EOP annexes that may be affected need to be reviewed prior to the drill by the planners to assess whether or not the annex procedures were followed.

Drill Moderators

After the type of incident has been chosen, additional decisions and personnel assignments may begin. Planners will need to locate moderators. As with the tabletop exercise, moderators will be given a timeline and will meet independent of the drill's planners. At their private meeting, they will determine when and what additional information and factors will be added to the drill. The planners will also require observers be in place. As with the tabletop scenario, observers will be seen and not heard during the actual exercise. They will simply be there to observe the decisions and actions of drill supervisors and participants. They will take notes, and be prepared to offer honest and frank evaluations of the drill, which will include what they felt was good, and what areas they feel need improving. The observers invited to participate should be chosen from across the emergency response spectrum and should represent local, county, and state agencies. As earlier noted, federal agencies should also be included, and they may offer a different viewpoint for planners to review after the completion of the drill. The moderators and observers should have no direct knowledge of those who will be participating in the drill. Their ability to focus on the task without feelings towards the participants will make the drill review far more relevant and will allow all issues noted to be reported.

Incident Management—Unified Command

Given that this scenario will be an event that turns into an incident, and since it is an annual event that has utilized a successful organizational template, the ICS structure should be in place already. For our purposes, we will utilize a unified command (UC) comprising law enforcement, fire, EMS, and OEM. All members of UC will be from Salem Township. No county or state responders will be assigned to the UC for the event. There may be other municipalities within the county which also have fireworks displays and county OEM presence is not required in Salem. The ICS structure will be established to the group/division level, and operations will consist of the following branches: law enforcement, fire, EMS, and public works. No staging areas have been designated. The branches will be broken down as well, and each branch will be outlined below. Law enforcement will include a four-division traffic assignment, patrol, bicycle unit, and a plain-clothes unit. The fire branch will include the suppression group, water supply group, and a fire watch group. The EMS Branch will include a three-location first aid station division, a transport group, an ALS group, and a bicycle unit. Public works will include a trash removal group, an emergency response group, and the maintenance group. All these overhead positions should be filled to provide a sense of realism. The actual number of responders working in each division or group can be modified and, if needed, they may be simulated as well.

True Nature of the Drill to Remain Secret

In order to truly test the abilities of the emergency responders, the nature of the drill needs to remain confidential. Providing street-level responders with a timeline of the drill's events will enable them to prepare responses to the emergencies prior to their occurring at the drill. Believing that the drill is to prepare them for the annual celebration will put them at ease; since many of the responders will have already participated in the celebration, they will have a certain comfort level. Once in the field and after the attack scenario is released, observers will be able to evaluate their responses. It would be desirable to have large numbers of volunteers on hand for the drill, but locating them may be difficult. There are several possible venues for soliciting volunteers. Boy and Girl Scout organizations may see this as an opportunity to get their groups badges for their volunteering, schools may be able to involve students and their parents, and even college students may be interested in becoming volunteers. The presence of volunteers will be important during the drill. Once the bombs are detonated, the volunteers can become victims. Each volunteer can be given an index card that will contain a description of his or her injures and wounds. The responders will then make decisions in triage, treatment, and even survivability. Responders on the scene for the drill can exercise their plans for a mass casualty incident (MCI), and UC and branch directors and other overhead responders will be graded on their response to the incident.

The decision to solicit volunteers will entail additional work for the planners. Their ability to gather volunteers to serve as victims will enhance the scenario by providing responders with people in need of their assistance. The planners will need to be careful not to divulge the true nature of the drill to participants, including the members of UC. If they believe that the drill is simply a review of the operational picture for the celebration, the actual drill will become even more realistic. Planners must have accurate numbers of victims and they must meet with either the volunteers or with representatives from the organization providing the volunteers. These volunteers need to know what to do, when to do it, and how to do it. Even if a small group is available, the volunteers can be given multiple roles, to maximize their participation.

Drill Communications

Another issue that will need attention will be ensuring that all responders are using the communications system that will be in place during the real celebrations. This will add realism to the event and will enable observers to assess the responders' communications abilities. During the events in New York City on 9/11, responders from New York City Fire and Police were unable to speak directly to each other. The federal government has since allocated tremendous amounts of money to various groups to develop interoperable communications systems in the aftermath of 9/11.

This drill will be an excellent opportunity for observers to assess the communications abilities of the participants. No special arrangements outside those made for the actual celebration need to be made.

To this point, we have chosen an incident, established an ICS structure, obtained moderators and observers, and made arrangements for volunteers to serve as victims. The next decision will be to determine how many responders will be actually on hand for the drill itself. There will be a need for police, fire, EMS and public works employees to be available during the drill. This may require overtime pay for members of these career departments or agencies that will be needed during the drill. Grant opportunities are available for training and these grants can offset most of the expenses generated by the drill. NIMS compliance recommends annual training exercises between response agencies and this drill satisfies that requirement. Using this information in the narrative of your grant application may help it to be received favorably, and it will stand a better chance of being selected for funding. Adequate numbers of responders from each response community will be required in order to get an accurate picture of how they will respond in a crisis. Once the adequate numbers of responders has been chosen, the next step will be to choose a date and time to conduct the drill. Remember, in order to get a true picture of response strategies and tactics to be used by both overhead and street responders, it is critical that they be unaware of the scenario that will unfold during the drill. Putting them in a comfort zone and then introducing chaos and an unexpected incident will allow for an excellent opportunity to observe how the responders think on their feet.

Sample Timeline for the Drill

Below is a sample timeline for the drill that will be developed. Note that the detonations are not included. As a result, the participants will be caught off guard when the detonations are introduced. As with the tabletop, the drill will not be conducted in real-time and participants will be so advised. The only real-time aspect of the drill will be the evaluation of the response to the detonations.

Timeline for Salem July Fourth Celebration

0700: Meet inside the Salem High School gymnasium for welcome and assignment designations.
0800: Participants report to their designated locations.
0845: Simulate band playing, and daytime activities.
0910: Report of lost child in the park.
0925: Report of possible heart attack in Grove #2.
0940: Report of stolen purse in the area near Grove #4.
1000: Simulate darkness, prepare for the fireworks display.
1020: End of drill.

As with the tabletop exercise, the true test of the responders will not be listed on the timeline. The reported detonations will come from the moderators located at the command post. Then, they will add additional information and facts as the drill unfolds. The secondary explosions will be announced, reports of fatalities and critically injured citizens will be broadcast, and field responders' responses will be noted. The victims can simply lie down, and present the index cards with their injuries to responding emergency personnel. Observers for review will chart how the responders react, their decisions, and requests they make at a later time. Responders participating in the drill will be lulled into believing that the simulation for the drill will end at 10:30 A.M. Nothing will be further from the truth, as moderators will begin infusing additional facts and events into the timeline in order to assess responders' reactions.

The drill will have to be conducted in a large area to simulate the incident happening during the annual celebration. If the park cannot be secured for the drill, alternate locations need to be located. The high school is an example of a large campus area where responders can be kept apart to simulate the distances in the park. The volunteer victims need to be split evenly into the areas where the explosive incidents are to occur. The reporting location can be placed on the index card containing the injuries that they will exhibit after the detonation. These types of drills are best conducted during warm weather, and obviously, on weekends. This will maximize the availability of volunteers to serve as victims and the parents of Boy and Girl Scouts may also be available to assist. Planners may also want to consider having refreshments served either before or after the drill is completed, for volunteers and responders.

Conducting the Drill

The drill can be conducted in four to six hours, depending on the scope of the incident and the number of volunteers and participants. The initial part of our selected drill can be covered very quickly and, when participants believe that the drill is ending, the real incident is revealed. This is what would happen to the responders if the incident discussed were to actually occur. The fireworks signal that the end of a long day was near and responders have a natural tendency to let down their attentiveness, believing that the celebration went very well. Once they enter this mindset, the incident presents them with unexpected challenges, which is would happen in the real world. The reaction time, the time between when something happens and when action is taken, will be tested here. Not only will their reaction times be tested, but also their judgment, their calmness or level of excitement, and their ability to focus on the crisis will be observed and recorded.

Moderators and observers will chart the response times of the responders in getting to areas where victims are located, and will also assess tactical decisions made by both rank and file as well as supervisors. How the responders adapt to the stress of an unexpected event will certainly be among the criteria listed by observers. Additionally, resources requested, the calmness or lack of cool in their voices will be noted, along with their organizational thought processes. Will they have thought to bring resources to the scene to search for secondary devices? Or will they have totally forgotten their training regarding these scenarios? Are the responders making reasonable requests for resources or are they totally irrational and unrealistic? These are areas where moderators will focus their energies. This, too, is a perfect example of testing cross-training of responders in areas outside of their normal responsibilities, such as first aid, triage and incident management principles. There will be a large number of victims and all responders on the location will be expected to conduct first aid, ensure the safety of the victims and survivors, and make resource requests to mitigate the emergency as quickly as possible. Fire and or EMS workers with terrorism awareness training may be aware of the potential of a secondary device and may also be sensitive to the possibility of a dirty bomb or chemical agent distribution scenario. The more responders trained the better, and there is then the distinct possibility that the critical indicators with this scenario will not be overlooked.

Drill Conclusion

Once the moderators and observers feel that they have injected enough scenario changes into the drill and have evaluated the responders' performances, the formal aspects of the drill can be concluded. The volunteers and participants can be summoned to a common area for a short debriefing. An informal review may be conducted at this time and the opinions of the responders and the victims can be solicited. Comments made by victims are often very insightful because participants are evaluated by untrained civilians who have a different mindset and level of expectations of the responders. Responder responses as well as those of the victims should be memorialized, because they will provide great insight as to what was going through their minds just prior to, and immediately after the unexpected detonations. These impromptu comments can be reviewed and dissected during the review that will be scheduled for a later date time and place. Moderators and observers will then get together and discuss their observations and the overall effectiveness of the drill's participants. They will begin their review process and prepare to discuss their findings at the formal review. An independent evaluation of their individual goals for the responders for each item interjected into the drill will be provided.

Review Period for Participants

The next step in the process will be for the planners to allow adequate time for themselves, observers and moderators to review and organize their notes. A formal review date needs to be established, and all participants must be given the option to attend. This may generate additional overtime expense, but all participants are entitled to be part of the process that rated their performances. This independent review will allow for the individual responders to make adjustments to their tactical decision-making processes. Suggestions from coworkers are often unheeded, simply because of rivalries or tension between members of the same agency. Many years ago, my regular partner was assigned to work with a different officer for a night shift. During the course of their shift, a diner was robbed at gunpoint. A local officer located and stopped the car and my partner came from the opposite direction. Upon arrival, he blocked the escape of the suspect vehicle by pulling in front of it at a forty-five degree angle, passenger side door blocking the driver's side front end of the suspect vehicle. He believed this to be a sound tactic. Once I pointed out that his partner was between the suspects and him, there was a disagreement. I felt that his partner was placed in the kill zone, should the suspects have desired to shoot their way out. We argued this point for months, with each of us believing we were correct. At a tactical training seminar a few months later, we approached the instructor with our scenario. He agreed that the position of the vehicle was improper and had needlessly exposed the partner to extreme danger. At this point, my partner accepted that his vehicle placement and tactics were improper. Since the review was from an "expert" rather than a colleague, he accepted the advice. When input is provided from an outsider who has no apparent agenda, responders are much more likely to have a positive response to the suggestions and to implement the strategy or tactic proposed to them.

The review should be conducted as was discussed in the last chapter. All participants should be afforded the opportunity to ask questions, and have certain strategies, tactics or decisions that were made explained. Not allowing these processes to take place will limit the effectiveness of the drill. The person determined to be the leader of the review must allow all participants to have their questions answered, but also must not allow any one responder to dwell on an issue for too long. The review must be presented in an orderly fashion and must be kept moving. Written reviews by all those in reviewing positions will help ensure that all areas of the drill were reviewed and will allow for a concise, organized review. A panel of reviewers may also be an option. In this instance, they are seated at a table and a round-robin discussion is conducted from the beginning of the drill to the end.

After-Drill Adjustments

The last step in this process will be to review any annexes or policies and procedures that were tested during the drill process. In most cases, there are several areas of policy and procedure that need to be adjusted. Some of these adjustments are easily made, while others will require more time and attention. Generally, issues brought to light during drills such as the ones in this chapter, require additional training for the responders. Sometimes it will be refresher training, yet other cases require that a new training class or technique be developed and delivered to the responders.

Whatever issue is raised, it will need to be addressed or the deficiency is liable to present itself again in a real-world scenario.

Additionally, there should have been several annexes of the local Emergency Operations Plan (EOP) that were applicable to the training scenario. These annexes will need to be reviewed and changes may be required of them as well. OEM personnel will then be tasked with conducting the updating or revising of the affected annexes. Once the policies and annexes are updated, all personnel affected by the changes need to be advised of the changes and new policies. As noted, this will sometimes require additional training for the responders. Once the needed training is identified, administrators must arrange for the training to be conducted and to assure that the trainer chosen is well qualified and aware of the new policy or direction of the response organization. The new policies or procedures cannot be implemented until all responders and affected staff has been trained on the revised or new protocols.

Summary

To summarize, the hands-on drill differs from the tabletop drill in several key areas. In the hands-on drill, there is far less simulation or modeling of positions that will be filled in a real incident. The overall cost to conduct this type of exercise will be significantly higher than that of a tabletop drill. In the tabletop exercise, the training emphasis is clearly for the supervisory personnel and the hands-on drill will test the supervisors as well as the responders who will be arriving at the incident. The number of participants will be greater in the hands-on drill and this type of exercise allows for independent analysis and evaluation of personnel by observers with no agenda or preconceived notions regarding policies, procedures, or the past performance or reputation of an agency's staff.

As with the tabletop scenario, the incident or event to be drilled must be determined by the planners. Generally, a drill is best served by using an incident and by allowing for something that may not have happened, or that the responders may have never contemplated. Drilling on events that occur on a regular basis will allow for a comfort level of the responders, but may not give an accurate picture of how your responders will react to an unknown. Even though that is something they do on a regular basis, adding unexpected events to the scenario will stress the responders and allow for a thorough review of their abilities and performance under simulated conditions. Accordingly, planners need to determine goals and objectives for the drill. There must be a skill set or process that planners are trying to evaluate and grade. The goals chosen may be very basic, or if the agency is well trained, there may be a need for a complex evaluation of personnel. Either way, goals and objectives will need to be developed and documented.

Moderators and observers need to be advised of the goals and objectives, but need not review departmental policies and procedures. If policies and procedures are reviewed, a more in-depth review will be possible as the moderators and observers will be familiar with what your agency expects of its responders. Also, knowledge of the policies may allow the implementation of a better list of information and scenario changes during the process of the hands-on drill.

After the scenario and goals are determined, observers and moderators will need to be chosen. These volunteers must be well trained and qualified since they will provide the independent review process for the exercise. Further, the moderators will be the participants that add the unexpected occurrences to this drill. They need to be familiar with goals and objectives and must maintain a level of secrecy throughout the process to ensure that participants remain unaware of the underlying issues prior to the drill. Having good moderators and observers can help to ensure that the drill is professionally and expertly developed and exercised. Additionally, these participants must have no internal or external ties or relationships with any participants or participating agencies. The farther removed they are from the response agencies, the better off the planners will be. These moderators and observers should be from across the response spectrum, and across the entire response community, from local, county, state, and even federal agencies. There is a definite need for these participants to be free from prejudice and pre-conceived notions about the participating agencies or individual responders. This anonymity will allow the reviewers to present honest feedback without the fear of insulting or agitating any responder who was at the exercise, and who may take umbrage at the review.

The next step in the process will be to schedule the drill location, date, and time. The timeline needs to be finalized, and all relevant, unsimulated positions need to be filled. Unlike the tabletop exercise, there will be more rank and file responders participating in this exercise. These responders will be used and evaluated by the moderators, and their reactions, response times and decisions will all be documented and become part of the review session. Having nonsupervisory responders participate in the drill affords observers a chance to determine the level of training or tactical decision making ability that the street responders possess. This review can assist the agency policy makers in revising the procedures and policies of their agency, as well as providing guidance on which areas that agency needs to enhance its training capabilities. Having this review done by an independent, unbiased, and qualified observer will truly provide a direction for future policy and training decisions.

After the actual drill is concluded, participants should be debriefed and their comments memorialized. These comments will be of great value to agency administrators, and along with recommendations of observers, will assist in policy and procedure updates.

A key component of the scenario in this chapter was adding an incident to an annually planned and executed event. This may have led to a sense of complacency on the part of the drill participants. This feeling that everything was under control was quickly and aggressively dismissed by the introduction of an explosive incident. Responders will have had time to review their own reactions and decisions prior to their exit interview, or debriefing. The purpose here will be to entice the responders to self-evaluate, and self-critique. This type of self-evaluation is sometimes helpful prior to a formal review, and responders' opinions of their performances can be verified, or brought to light by the reviewing observers. Serious minded responders will often find themselves second-guessing their decisions and actions during the incident. This will make them even more attentive during any future review of the drill.

The next step in this process is to conduct a formal review. This will be scheduled for a time after reviewers have had time to review notes and to prepare formal evaluations. These evaluations, as noted earlier, need to be honest, and frank. Nothing, good or bad can be held back and, in this manner, planners and agency administrators will have a clear picture of the policies and procedures that need to be addressed. All participants should be afforded the opportunity to be present and all of their questions should be addressed, without the moderator of the review losing control. The review may be conducted in a short period of time or it may require several hours or more. The duration of the review will be contingent upon the scope and complexity of the exercise, the number of participants, the success or failure of the stated goals, and the number of questions presented to the observers. Regardless of the time required to complete the review, it must be completed to assure that the observers presented a clear operational picture of the drill to agency administrators. Participants must be reminded that the review is not a personal attack if they are the subject of an action that is in need of correction. Egos must not be allowed to enter into the review process. Any deficiencies in the drill that are not addressed can have serious negative consequences. If poor decisions and tactics are left unaddressed, those who made the errors will believe that their performances were satisfactory. This will reinforce a poor decision or tactic in the mind of the participant. Obviously, in a real event, these decisions could have dire effects.

Once the review process has run its course, the next and final stage of the exercise will require a meeting of agency administrators. Observers, moderators, supervisors, and responders who participated in the drill should have outlined and detailed several areas in their agency's policies and procedures that will need to be evaluated and updated or revised. The same may be required of any EOP annex that was subject to review during the exercise. The policies and procedures or annexes must be thoroughly reviewed and any proposed or recommended changes need to be seriously considered. Changes will then be made and arrangements for training on the new policies must be undertaken. A qualified trainer with knowledge of the intent of the new policy will be needed. Once all members of the agency have been so trained, the new or revised policy or procedure can be implemented.

Hands-on drills offer something much different from tabletop exercises. The ability to observe and exercise more members of your agency will afford an ability to truly see how your members operate, and in which areas where there is a demonstrated need for additional training. A cost factor may be involved in this process, but in the final analysis, the money spent may be well worth it for your agency. If your agency is truly unable to afford the costs that will probably be incurred to conduct the training, you have several options. Your first option would be to engage other agencies into the drill, thereby reducing the number of your own personnel required to attend. This process should only involve other agencies that would normally be called in to assist your agency in a real incident. Asking one or more sister agencies to train will also provide additional benefits to you. The responders who are participating will see how members from another agency react to the same set of facts as those given to your responders. This exercise will afford them the opportunity to train together and they will begin to gain a level of trust in each other, and possibly they will begin to recognize the skill sets of their peers. This will be especially beneficial regarding your supervisory personnel, who may be in a position to someday direct the resources from another agency. Working together can assist in breaking down some of the walls that were discussed earlier in this book. The second option for funding your exercise would be to apply for training grant monies. Online search engines may be able to direct you to these grant opportunities. If you decide to go this route, review the grant application tips presented in Chapter 10.

Hands-on drills take time to plan and execute, but their value to an agency can be immense. Do not shy away from these exercises because of the amount of preparation required for a successful scenario. You will be glad you exercised your agency if disaster strikes and your responders all make good decisions, especially if they were following revised policies amended after the drill. Good leaders and those wishing to be proactive will see the benefits to hands-on training, and make every effort to provide this type of training for their responders.

Chapter 14

Putting It All Together

Emergency responders of all types have always had difficult jobs, often completed under trying, fast-moving circumstances and with the potential for the loss of the lives or property of the citizens they have sworn to protect. The response community has faithfully carried out these difficult assignments with compassion, professionalism, and a sense of pride. Their duties are constantly changing and new challenges are presented on a regular basis. Responders must maintain advanced levels of training, and must be ready to accept and manage new challenges and training courses as they arise. The positions remain the same, titles and overall responsibilities remain unchanged, but there are subtle nuances of discipline that each responder must address. If a study of emergency response was to be conducted and the roles, duties, and equipment issued and utilized were cataloged, the study would reveal significant changes from generation to generation in each field of response.

A Century of Change for Law Enforcement

Law enforcement has seen its share of changes in the past one hundred years of its history. Patrol duties have been enhanced and made easier by the introduction of new technology. Cruisers and motorcycles were rare or non-existent for patrol officers just over one hundred years ago. Most police officers were designated to a foot patrol area, often referred to as a *beat*. The officers walked the streets in their foot patrol area and got to meet most of the residents. With the advent of motorized vehicles, the interpersonal contact was reduced, but the benefits were faster response times and a greater area of a city or town receiving police coverage. The *call box*, a staple of police communications was replaced by mobile radios, and eventually portable handheld radios. Again, technology was there to enhance the job productivity of the officers. Between the radio dispatch and the cruiser, law enforcement was able to provide much faster response to all calls. Dispatch centers were able to more closely monitor their officers, and as a result, officer safety was greatly enhanced. Communication was immediately available between supervisors and other incoming officers at incidents, which increased overall efficiency for the departments and provided better coverage in the response areas.

Better Weapons and Weapon Systems

In the early part of the twentieth century, law enforcement officers were totally outgunned by the criminals. The criminals were using assault rifles and Thompson sub-machine guns, while our responders were forced to carry six-shot revolvers. While law enforcement officers continued to be gunned down, semi-automatic handguns eventually were made available to law enforcement to try to make things slightly more even between police and criminals. More and more violent and sophisticated criminals and crimes resulted in the law enforcement community developing special response teams, most frequently referred to as special weapons and tactics (SWAT) teams. These officers were trained in advanced weapons systems as well as strategies and tactics to meet these violent criminals head-on. Additionally, technology was available to the law enforcement officers in the form of soft body armor, also referred to as *bulletproof vests*. These vests are not actually bullet proof, but are designed to stop projectiles from handguns, shotgun pellets (not shotgun slugs), and some slow moving rifle rounds. The ballistic panels used in these vests are not designed to protect officers from knife attacks, specifically stabbing injuries, although they will provide some protection against slashing attacks.

Even with this technology, officers needed still more tools to safely complete their missions. Police were instructed to refrain from deadly force unless specific criteria were met and the threat was imminent. Protection was needed for violent criminals who were unarmed, but were physically superior to the responding officers. Tools developed include mace, oleo capsicum (OC) spray, straight batons, side handle batons, collapsible batons, tazers, and stun guns. Older officers were known to carry blackjacks and *slappers* designed to subdue unruly or resistant subjects. Even additional tools were invented to assist them. Police use radar to determine the speed of vehicles to enforce speed limits and in-car video systems were developed to enable officers to document encounters with the public and to document conduct of both the officers and the citizens. Blood alcohol detection technologies have been developed to assist or confirm that drivers are operating their vehicles at or above the legal limit for blood alcohol concentration.

Air Support and Related Technology

Larger cities and counties employ helicopters to cover large expanses and to assist ground units in the apprehension of criminals and aggressive drivers. Forward

Looking Infra Red (FLIR) technology and Thermal Imaging Cameras (TIC) have been developed to help officers see in the dark and to track subjects. Some law enforcement officers now carry radiological detection pagers to detect radiation levels which are above normal background readings in their areas. Airports and courthouses have x-ray machines and metal detectors to assist in preventing the introduction of weapons inside their buildings. Police cruisers have computers in the cars to assist in completing reports more quickly, verifying active warrants, and completing motor vehicle registration and driver license checks. Some agencies have issued handheld computer devices for field use.

Even after reviewing all of the progress that has been made for the street officers, they are still lacking in many areas. Training requirements and mandated classes and refresher training consume a tremendous amount of work time each year. The advent of terrorism and terrorism response and detection has added to the already full plate of responsibilities of our law enforcement officers and police and sheriff personnel are now forced to expand their roles yet again. They are being asked to do more for us, but are given less time and inadequate training for this new mission. In fact, it has been shown that these officers are lacking in other skill sets that would make their jobs safer and also make the jobs of other responders safe as well. These areas have all been addressed earlier in this book.

Fire Service Improvements in the Last 100 Years

The fire service has also seen tremendous changes in the last one hundred years as well. In many parts of the country fire hydrants provide a ready, accessible water source for firefighters to access for suppression of fires. There are still areas where there are no hydrants, but usually there is a water delivery system present where large numbers of citizens live. The ability to get water onto a burning structure quickly is obviously of paramount importance and firefighters of yesteryear did not always have that availability. Additionally, the water pressure in many areas has been substantially upgraded, which will further enhance the ability of firefighters to suppress large fires.

Better Apparatus

The apparatus that firefighters use has also been greatly upgraded over the many years. Fire engines or pumpers now have the ability to pump at volumes exceeding 2,000 gallons per minute. The fire service has pumpers with onboard tanks that can exceed 3,000 gallons, and other pumpers have tanks that carry as little as 100 gallons. Fire engines or pumpers are now much safer than was the case even a generation ago. Specialized equipment such as brush trucks, or wildland interface vehicles allow fire crews to drive into wooded areas to track and extinguish brush fires. Fixed wing aircraft and helicopters also are used to combat these wildland fires that consume millions of acres every year. Regulations now require that no firefighters respond to fire calls while standing on the back step of the engine so all new fire apparatus must be designed and manufactured to ensure that all responders can be seated inside a cab, much like cars. Crews are required to wear seatbelts and no responders are to stand while responding to calls. A recent change mandates that fire helmets cannot be worn in apparatus due to their shape and possibility of neck injury after a collision. Further new standards mandate reflective striping similar to sergeants' chevrons be applied to the rear of all newly manufactured fire apparatus to ensure visibility by oncoming drivers.

Most new vehicles are equipped with heaters and air conditioning units for both the front and rear passengers, and thus the vehicle can serve as an area for rehabilitation. Firefighters can retreat to their apparatus to escape temperature extremes and take their breaks in a more suitable environment. Some fire apparatus also have computer systems on board for the responders and these are of great assistance. They can display fire hydrant locations and fire personnel can install fire pre-plans for certain locations and can have hazardous materials reference texts installed as well. The pre-plans serve as a response template and include information that will make the operation safer for fire crews. This information may include utility locations, owner and alarm information, construction types, hydrant locations, hazardous materials present, and special concerns.

Suppression Technologies

Fire engines now can carry different types of fire fighting foams for various fire suppression applications. The engines also may be equipped with the foam deployment tools already installed on the engine, or they may be designed for use closer to the fire or spill. Foam systems are designed for ordinary or Class A fires such as wood, paper, furniture, etc. Foam is also available for Class B fires which include flammable liquids such as hydrocarbons (fuels) and alcohols. New technology now includes the Compressed Air Foam System (CAFS), a system that utilizes compressed air to generate the foam, thus using less water. The advantage here is that the fire hose is lighter and easier for the fire crews to maneuver as compared to a hose with all water.

Improvements to Aerial Devices

Fire trucks, ladders, or aerials have also been improved to enhance firefighter safety. Fire trucks are now equipped with aluminum or steel ladders that can reach heights of more than one hundred feet. These trucks also have sophisticated hydraulic systems for raising and lowering the aerial device. The outriggers, the arms that extend from the sides of the truck to provide a stable base for the aerial to be raised, are also hydraulically powered and provide a much safer platform from which to operate these devices. Technology also provides safety features to prevent the truck from tipping over while the aerial devise is deployed and also to assist the operator in *bedding* or replacing the ladder into the stowed position without damaging the apparatus.

Other technologies include pre-piped waterways for the aerial devices so that the firefighters need not bring a hose line to the tip of the aerial to flow water. Large pipes are connected to the ladder and operators need only establish a water supply to get water to the tip of the ladder. These aerials may have two nozzles or *monitors* on certain platforms and these can flow more than 1,000 gallons per minute each in either a fog or straight stream pattern, as determined by the operator. Truck companies also have platforms or *buckets* atop the aerial devices. These buckets are small, enclosed areas where two firefighters can stand to either effectuate rescue, make ventilation cuts on roofs, or direct fire streams. Technology also allows firefighters to control the movement of the bucket from inside the bucket, rather than relying on the operator at the turntable or base of the truck. These platforms may have a ladder attached and are called a *tower ladder*, or they may not have a ladder, in which case they can be called a tower. Both ladders and platforms may also have a large cylinder of air affixed to the aerial to allow for extended operating times in atmospheres requiring the use of SCBA.

Ventilation Improvements

Other innovations for truck company crews include power saws for ventilation as well as for forcible entry situations; these are vastly superior to axes which are still in use. Gasoline powered *positive pressure ventilation (PPV)* fans and electric powered *smoke ejectors* or negative pressure fans provide multiple options to assist firefighters in ventilating, or removing smoke or toxins from fires. Hydraulic tools such as *rabbit tools* and similar forcible entry tools make the task of gaining entry through locked doors much faster and less labor intensive.

Air monitoring technologies such as multi-gas and single gas detectors also have made fire-fighting operations safer. These devices monitor the levels of certain gases that may be present at structure fires, in sewers and other confined spaces. They also detect carbon monoxide releases in residences. Carbon monoxide is an odorless, colorless gas and without meters, detection cannot be made. Exposure to certain levels of carbon monoxide can prove fatal and every year, many people are killed by releases of this gas inside their homes. Additionally, these meters can alert firefighters to the level of oxygen and other gases present in certain applications. Thermal imaging cameras have also been introduced to allow firefighters to *"see thru the smoke"* and help in search and rescue applications. This technology allows search crews to complete their searches much more quickly, and is also used to locate missing or trapped firefighters.

SCBA and PASS Devices

The gear the firefighters wear has also undergone a tremendous upgrade. In years gone by, firefighters wore rubber coats and boots that were pulled up to about mid-thigh.

Currently, fire jackets and pants, or *bunker pants,* are made of nomex fibers. They provide much greater protection in high heat environments. Nomex hoods protect the neck and ears of the firefighter. Self Contained Breathing Apparatus (SCBA) have been developed and upgraded in the last few generations. The cylinders are now made of composite fibers and lightweight materials that make wearing them less taxing. This technology allows firefighters to go deeper into burning buildings in order to effectuate suppression and rescue efforts. The newer SCBA are also equipped with *Heads Up Display (HUD)* systems that allow the wearer to see how much air remains in their cylinder. This technology is relatively new, but has played an important role in ensuring that firefighters have enough air in the SCBA to allow them to safely exit the fire. Also relatively new is the *Rapid Intervention Connection (RIC),* which allows a downed or trapped firefighter to receive another cylinder of air. The cylinder is delivered to the down firefighter and connected to his SCBA through a designated connection. The cylinder is opened and the firefighter has more air. This new connection removes the need for the firefighter to remove the mask, or the entire SCBA to get more air in an emergency. Another technological advance is in the *Personal Accountability Safety Systems (PASS),* which has been developed to track firefighters at an incident. The firefighter is required to wear the unit on their gear. If he stops moving for a designated period of time, a warning alarm sounds and grows louder until either the firefighter moves or it reaches its loudest point. The alarm is loud enough to enable searchers to locate the downed firefighter without the firefighter needing to be in contact with them. Technology now has provided that this PASS can be built directly into the SCBA, and once the air cylinder is opened, the PASS is armed and ready to protect the user. Other new technology includes devices to determine if an electrical system is energized or not, which can prevent injuries and death due to electric shock.

As with the law enforcement officers, technology has been kind to the fire service. But, also as with law enforcement responders, mandated training and time required to maintaining these high-tech devices has prevented firefighters from getting the training that they may need in cases of terrorist attack. There are many areas where training can be enhanced and the fire service needs to take advantage of these training opportunities as well.

EMS Improvements over the Last 100 Years

The Emergency Medical Service (EMS) has also benefited from technology in the past century. The ambulances themselves have evolved greatly. Radios, built-in oxygen delivery systems, and larger work areas are but a few areas where the service has been enhanced. Lighter weight oxygen cylinders, lighter and more stable patient stretchers, and lightweight back boards have been introduced. Semi Automatic External Defibrillators (SAED) are now prevalent in airports, malls, schools and other public locations. They monitor the heart for the two shockable rhythms—ventricular fibrillation and ventricular tachycardia. The SAED analyzes the heart and advises rescuers as to whether or not a "shock" is required for the patient. If warranted, the defibrillator, having been pre-set, will advise when it is charged and will deliver the appropriate number of "*joules*" to the patient.

CPR Evolves

Cardio Pulmonary Resuscitation (CPR) has also been studied, and the rate of breaths to compressions has undergone many changes over the years. There are now machines that will perform chest compressions for the rescuer. Doctors in hospitals can now relay orders to EMTs and paramedics in the field can now send telemetry or analyses of the heart over radio waves for examination. Advances in medication and medical delivery systems have made EMTs and paramedics more efficient. Development of treatments for seizures, asthma attacks, and insulin for diabetics can be delivered in the field and have saved many lives. Helicopters are used to transport critically injured victims from the incident locations directly to hospitals and trauma centers. The travel time via helicopter is superior to ground transportation and the time saved in transit has saved many lives. Other modifications to blood pressure equipment, re-usable soft splints, and traction devices have all helped make the jobs of the medical responders faster and easier. But, as with law enforcement and fire service responders, the training and re-training and re-certification processes have eliminated the ability of the EMS community to get training which crosses their traditional boundaries and this needs to change as well.

Hazmat for EMS

One hundred years ago, the term hazardous materials team was unknown. Any incident involving what is now deemed to be a hazardous material or substance was handled by the fire service. Much like a paramedic, or a SWAT officer, the hazardous materials technician or specialist is an offshoot of the fire service. Since this specialty has been implemented, technology has assisted them greatly in performing their tasks. Obviously the hazmat responders wear SCBA, but they also have the choice of three levels of chemical protection. For the most serious incidents, the hazmat technician will wear Level A, fully encapsulated suits, complete with SCBA. This allows entry into the harshest Immediately Dangerous to Life and Health (IDLH) atmospheres. Firefighters have multi-gas meters and the hazmat teams use these same tools. However, the hazmat teams also have the ability to identify liquids and other gases and to monitor the atmosphere for many more gases and vapors than the fire service. Many tools are now available to assist these responders in the identification of many hazardous materials and substances. Hazmat technicians can detect radiation and also identify many radioactive isotopes with the latest technology. Hazmat teams also have the ability to field test for biological warfare agents. In addition, these teams can bring portable weather stations into the field and these can be of great assistance to incident managers who need instantaneous weather conditions at the site. Wind speed and direction and relative humidity can have great impact on certain types of incidents such as chemical spills and releases. These teams also carry tools and equipment to clean areas after the release of liquids such as oil, gasoline, diesel, kerosene, and other fluids released in automobiles accidents.

Hazardous Materials Teams—Decontamination Types

Hazardous materials teams also specialize in the different types of decontamination including definitive, technical, self, and emergency. Hazmat teams are also trained in decontamination techniques for exposure to radioactive particles. This type of decontamination requires a *survey meter* and it is a very slow, time-consuming technique. The person exposed stands still while the meter locates the radioactive particle. Once located, duct tape, or a similar tool is used to remove the particle from the victim. Self-decontamination is when a rescuer is able to complete the decontamination process on his or her own, generally by passing through a decontamination shower. Technical or definitive decontamination results when all contaminants have been completely removed from a person or equipment. The tool is then safely placed back in service for use at future incidents. Emergency decontamination results when a victim is in need of immediate medical care, but must be decontaminated prior to treatment. The victim is quickly exposed to copious amounts of water, stripped of contaminated garments, and then turned over to the medical responders. Decontamination requires a large area, a water supply usually provided by a fire engine, and a device to collect the used water, often called *runoff.* The runoff or contaminated water is collected and placed in fifty-five gallon drums, treated as a hazardous material, and removed accordingly.

Control Zones

Finally, hazmat teams establish and delineate the three control zones at an incident. The three zones include the hot zone, the warm zone, and the cold zone. The hot zone is the area of the release of a product. The warm zone or contamination reduction corridor is where responders dress, undress and undergo decontamination. The cold zone is an area outside the hazard area, where it is safe for anyone to be without any protective equipment. Hazmat determines these zones, and they are strictly enforced for all responders. Hazmat teams are routinely asked to respond to various odor calls and releases. These responders are the ones designated for product identification and purity. But, as with the other response entities, there is not enough time in the work schedule to allow training in other responders' realms. This, too, needs to be addressed.

The emergency services communities have come a very long way in the past century. Yet even with all of the technological advances made in each response area, tremendous training deficiencies remain in all areas of first response. These deficiencies have been identified and explained, and recommendations have been offered for addressing these deficiencies. Training curricula and possible training locations and instructors have also been put forth. Tabletop and hands-on drills have been detailed and laid out for responders to utilize. But specifics need to be reviewed and options discussed, and re-examined.

Law Enforcement's Roles at Incidents

In the first responder community, law enforcement has the biggest burden because they will most likely be the first responding emergency services workers. Police officers and deputy sheriffs are on duty around the clock and are mobile. They work inside a mobile office and that office is taken with them to their incidents. Upon arrival, they are immediately expected to take definitive and decisive action to mitigate whatever incident they have been dispatched to. These dispatches certainly cross traditional responder boundaries. Let's take a moment to review exactly what law enforcement personnel are trained to do at each type of response. We will also examine the steps that are actually taken by law enforcement, regardless of their level of training. We will then examine areas where their response can be enhanced, what training will effectuate this response, and what specific equipment they should be exposed to.

Law Enforcement's Response to Fires

Police are dispatched to a structure fire in a house in their response area. They respond immediately and are on location BEFORE the fire department has left the station. The police officers can see visible flames and heavy smoke. They are trained to report a working fire at the location to their dispatch center, which relays this information to incoming fire units. Police should attempt to rouse any victims who may be in the home and advise them of the fire conditions. Law enforcement personnel are not trained in search patterns for victim location, nor are they trained in victim removal techniques. They also are not trained in ventilation techniques, or the importance of coordinating ventilation efforts with suppression efforts. Nonetheless, if a neighbor or resident advises them that there *may* be a victim in the fire, most officers will attempt entry. Given that their uniforms will not burn, but melt to them when exposed to certain fire temperatures, the decision to enter may not be wise. Additionally, given the fact that they do not have SCBA to allow them to breathe in the fire, or thermal imaging cameras to assist in locating victims, again entry would seem unwise. Lastly, law enforcement are trained to react to visual cues, and in a fire, the fire crew react not only to visual cues, but are trained to rely on physical contact with a wall, and verbal communication with other crew members. If suppression is your duty, contact with the hose line is the order of the day.

A law enforcement officer untrained in the effects of air on fires may very well be the cause of the fire growing in intensity and may also have the undesired effect of being trapped inside the residence. Yet, given all of these factors, the officer may very well attempt entry. Upon arrival at the fire, the officer's attention is drawn to the fire and he may park the cruiser in front of a fire hydrant. Another possibility is that the patrol car is left directly in front of the house, again obstructing the tactical response and staging locations of fire units. Given all of the areas where the officer is untrained, it would seem that the best option is to remain outside the fire and continue to provide updated information to the fire department. But is there any training that can be quickly given to the officers to allow them to do something without getting injured or killed? There most certainly is. The officers can be exposed to a brief class on the science of fire and the potential dangers faced by those without protective ensembles. Advising the officers where to place their cruisers to avoid disrupting tactical operations for fire personnel could also be explained. During fire department training, police officers should be invited to observe, and ask questions. Law enforcement officers should be exposed to equipment that the fire service will bring to calls for service including the air monitoring meters, thermal imaging cameras, fire extinguishers, and ventilation fans. The fire personnel should explain their priorities to the police, and also explain their capabilities and limitations.

Television and movies do a tremendous disservice to first responders. In the case of the fire service, untrained people may believe that one can enter a burning building with no breathing apparatus or fire ensemble. They believe that they can breathe in the burning building and that they will have excellent visibility for search and rescue applications. Lastly, they are led to believe that they can walk upright. These factors are one hundred percent Hollywood, zero percent fact. If law enforcement officers buy into these myths, they, too, will become victims for fire service personnel to rescue. A little bit of training for these responders can prevent tragedy at fire scenes.

Law Enforcement's Response to Medical Calls for Service

Law enforcement has few tactical duties at fires, but what about medical calls for service? Now we dispatch them to a drowning victim in cardiac arrest at a residence. The police arrive prior to the ambulance and report a male victim in cardiac arrest. He was found out of the pool on arrival. Law enforcement officers generally have a Semi Automatic External Defibrillator (SAED) and a first aid kit, possibly including oxygen. The officer begins a patient assessment and determines that CPR is required. Until EMS arrives to relieve him, nothing else can be done. The SAED can be applied to the victim for heart rhythm analysis, but CPR will probably remain the officer's only option. The cruiser is not set up for patient transport, nor can telemetry be sent to any doctor at a hospital. The officer must await the arrival of EMS, and surrender the scene to them. The officer will not attempt any advanced medical procedure, nor will he attempt transport. The police are very limited in first aid response. Additional training will be of little use in this case, but the officer should be trained in CPR.

Law Enforcement Roles at Hazmat Calls

Moving on, we will address a hazmat response for the officer. Now police are summoned to a motor vehicle accident on a two-lane rural highway involving a tanker truck and a passenger car. Again, law enforcement will arrive first with fire and EMS lagging behind. Traffic will begin to back up in both directions. There may be a leak of some kind from the bottom of the truck. Most police officers will bypass the traffic by driving on the shoulder of the road. Very little attention is paid to street elevation and wind direction. The officer gets as close as possible to the accident and exits his cruiser. But is this the proper course of action and is it safe for the officer? The answer is no to both questions. The best course of action would be to identify the product the truck was hauling, and determine whether it is safe to get closer. But don't police recruits receive this training during their basic training at the police academy? They most likely did, but accidents of this nature are not a common occurrence and if the material is not reviewed, the skill may be lost. Accordingly, hazmat refresher training should be conducted annually to assure that officers are familiar with protocols and procedures for hazmat response. The hazmat team members can provide the refresher training that would be required of law enforcement. The North American Emergency Response Guidebook (NAERG) is a tool that most cruisers carry, but over time, the officers become unfamiliar with the book and forget how to use the information contained in it. If we are going to dispatch officers to hazmat scenes, we need to be certain they are trained.

Law Enforcement Roles at Terrorist Events

Lastly, regarding police officers, we will discuss a terrorist event. Many police agencies took advantage of equipment purchasing opportunities provided by the federal government in the aftermath of 9/11. Various types of equipment and gear were purchased including Air Purifying Respirators (APR). However, very few law enforcement agencies went the extra mile to purchase chemical protective ensembles for their personnel. Additionally, given the need for annual training and fit testing for the APR, most agencies are no longer compliant with the OSHA Respiratory Protection Standard 1910.134. If there is a large-scale terrorist attack on our soil again, law enforcement will be first to arrive. Much like at the fire scene, the officers will have an overwhelming desire to begin helping victims of the attack. Survivors will greet the first arriving units and demand that the injured victims be assisted. Most officers will begin medical treatment. Officers will most likely be acting instinctively, and that will entail victim treatment and assessment. Preservation of life is always a top operational priority. Hopefully, additional resources are incoming, and an incident command structure will eventually be put in place.

Given that this response is to an act of terrorism, other tactical decisions need to be made. Arrangements need to be made for a security sweep of the immediate area and a check initiated for secondary devices such as bombs or other explosive materials. A decontamination line may need to be established that will allow victims to be placed into ambulances for transport for definitive medical care. Information regarding what has occurred, number of victims, types of injuries, and other observations will need to be forwarded to other incoming response units. The initial responders will also need to determine where vehicles need to be staged and assess the numbers of each resource that will be required to mitigate the incident as quickly as possible. Local, county, and most likely state OEM personnel will be enroute to assist the incident managers with ordering resources. Federal agencies will also be dispatched to the scene and these agencies will have their personnel integrated into a unified command structure upon arrival.

Given the nature of the terrorist attack and the response to it, having taken an incident command class would have been extremely beneficial to initial police responders arriving at the incident. The pressure put on first arriving response units will be overwhelming, and as noted, there are essential decisions and tactical operations to be addressed. In times of stress such as these, officers rely intuitively on the training they received to handle such an event. Should the responders have no training or experiences to draw upon, time may be lost while the officer is determining what actions need to be taken and in what order. Additionally, other factors such as current terrorist attack scenarios and methodologies will need to be evaluated to determine what other actions will be needed.

Having well-trained officers arrive immediately after an incident will be critical to the most efficient resolution to the incident. Having personnel who have been cross-trained will be beneficial in these responses. Officers who can determine which resources and the capabilities and limitations of those resources will make a huge difference in the possible outcome of these types of incidents. Since only the terrorists know the date, time, hour, type, and methodology of the next attack, all responders need to be prepared. We cannot allow complacency to prevent us from training our law enforcement officers to be as prepared as possible to serve as our front line of defense and response.

Law enforcement officers will generally be the first professionals to arrive at almost all emergencies. Of the four major response groups, law enforcement has the smallest "toolbox." Most of the tools they rely upon are carried on their belts such as firearms, spare ammunition, batons, pepper sprays, handcuffs and portable radios. Their cruisers may carry flares, first aid equipment and either a shotgun or rifle for certain emergencies. Other first responders bring more equipment to the incident than the police, and generally have some cross-training experience. The hazardous materials technicians and firefighters will have the broadest base of training and EMS will also have some cross-training skills needed for terrorist attacks. The scenario reviews for these responders will be considerably shorter because these responders are not sent to every type of emergency. Law enforcement is dispatched to every call that the other response communities are sent to; therefore, their level of cross-training needs to be the most comprehensive.

Fire and EMS responders are not sent to burglar alarm activations or traffic lights that are not working. Only law enforcement is dispatched to these service calls and unless a further investigation determines a special need, only the police will respond. Whereas law enforcement officers universally respond to emergency calls, other responders are generally called to respond only to calls in their primary field of training. The major incident type that will require these responders to cross traditional boundaries will be terrorism response. Their responses to acts of terrorism and the areas needed for cross-training each discipline will now be reviewed.

Firefighters' Roles at Incidents

Firefighters may be the responders with the broadest base of training, and they certainly have the biggest toolbox. They can be considered the jack of all trades, and they have training that touches on every emergency response discipline. They are trained to preserve fire scenes for possible arson investigations and can assist in traffic control. Firefighters generally have some first aid and CPR training and many fire apparatus currently carry SAED. With regards to hazardous materials, the fire department will usually arrive prior to the hazardous materials team. Firefighters are trained in defensive tactics for hazmat incidents and have basic equipment to meet this objective. These responders also have had some training via hazmat in responding to terror attacks. However, the training in this field is limited and needs to be reinforced and broadened.

Firefighters' Roles at Terrorism Events

Given the world climate regarding terrorism, all emergency responders should be aware of the potential for secondary attacks. Current trends even indicate a need to be aware of a tertiary or a third wave of n attack. Realizing that secondary attacks are a possibility is not enough. Fire service personnel need better training, which should include the type of attack expected, when, where, and how to prevent it. Taking a class on CBRNE or hazardous materials response once in a career will not be enough. Training courses on terrorist trends, suicide bomber awareness, and improvised explosive device (IED) awareness should be provided. The sad truth is that the material is available, but not presented to all responders. Fire crews have certain responsibilities during response to an act of terrorism. As with most human beings, they focus on what their training has prepared them to handle. As with the police, if they have no experiences to draw upon, they will not be totally effective. Fire and hazmat crews may be called upon for victim rescue after an incident, depending on the nature of the attack and the agent that was used. They will focus on their jobs, but they will have unanswered questions, such as those involving secondary attacks. All responders need to be fully prepared to render comfort and aid to victims, and in order to go "into battle" fully concentrating on their job, they need training.

EMS Roles at Incidents

The EMS community will have the most critical response role following an act of terrorism, especially if there are multiple victims. Think back to the Tokyo subway attacks of 1995 where there were thousands of victims. EMS will be required to assess, treat, stabilize, and transport many of these victims to hospitals. Once the operation stabilizes, victims will be brought to the EMS responders. These victims will need to have completed decontamination procedures prior to their assessment and treatment. The EMS response community will be almost immediately thoroughly overwhelmed. Their members will be part of a unified command structure once the resources to establish a proper incident management structure arrive. The location that is chosen to treat the victims will be of paramount importance. The purpose of secondary and tertiary attacks is to kill or injure the emergency responders. This goal will effectively amplify the total number of casualties at the incident. Injured people with no treatment are likely to succumb to their injuries. Attacking the responders that are trying to save lives will have the double goal of killing the responders and increasing the death toll. The location chosen for treatment of victims needs to be carefully considered and a security sweep will be required.

EMS responders should receive additional training in the areas of signs and symptoms of exposure to the various chemical warfare agents (CWA), including vesicants. Further, their training should be expanded to include current terrorist attack methodologies and trends. In Iraq, for example, terrorists and or insurgents have attempted to use chlorine trucks as IEDs. These attacks were widely unsuccessful, as the terrorists failed to determine the proper amount of explosive required to release the chlorine without incinerating it upon release. EMS responders en route to a CWA attack can begin to prepare themselves for what may await them, provided that they are aware of all of the chemicals and agents currently being deployed by the various terror groups. Accordingly, the additional training for them should include terrorism trends and current attack scenarios.

EMS personnel should also be trained in suicide bomber awareness. A theme of secondary attacks is to deploy a suicide bomber dressed similarly to he first responders, which allows the terrorist access to secure areas where responders are at work. Once inside the area, the suicide bomber is free to detonate an explosive device where it is deemed to be most effective. EMS personnel should be aware of this trend and be vigilant in monitoring their surroundings.

Hazmat Teams and Their Roles at Incidents

The last group of emergency responders reviewed is the hazardous materials technicians and specialists. As has already been noted, these professionals only respond to calls for service in their response discipline. Hazardous materials teams are a specialized resource and have very specific roles and duties at incidents. Should a terror attack contain biological or chemical agents, the hazardous materials teams will be deployed. There are very few teams and given the scope of the attack, they may be needed at multiple locations. The primary functions of a hazardous materials team at a terror attack will be air monitoring, product identification, and decontamination of victims. Much like the EMS community, the hazardous materials team will have many duties and not enough crew members to accommodate all requests.

In reviewing the responsibilities of the hazardous materials team, it is difficult to eliminate any of their primary duties. Incident managers and EMS personnel will need to know what agent was used and that agent's purity. The quality of the air must be monitored as well to ensure that responders can safely conduct their operations. A secondary attack with additional chemical agents, for example, could have a devastating effect on the operations that are underway. As noted earlier, victims of chemical attack cannot be treated or transported until after they have been through the decontamination process. A logical review of the three primary functions will indicate that all have life safety implications.

The hazardous materials team members will be stretched thin and will be focusing on their special roles in the mitigation process. Their training gives them information on the various agents that may be used in all types of CBRNE attacks. Hazardous materials teams can detect CWA, including vesicants, and can also field test for the presence of biological materials. If biological materials are determined to be present, further analysis may be required in a laboratory environment. Should the attack scenario involve radiation, such as detonation of a dirty bomb, hazardous materials teams can also detect the presence of radiation. The duties of these responders will be critical in assisting incident managers to plan and react to the issues faced after the attack. Identifying the agents used is very important to those trying to mitigate the incident as well as medical professionals trying to save lives.

The hazardous materials response community is well versed in responding to incidents involving the release of hazardous materials and substances. By definition, a terror attack using chemical, biological or radiological materials is a hazardous materials release. While these responders have a broad training base to begin, there are a few areas where their training can be enhanced. The additional training will be similar to the fire and emergency medical responders. The areas of suicide bomber awareness and current trends in terrorism, including attack parameters and dissemination methods of terror groups should be taught. These areas need to be refreshed on an as needed basis for all of the responders who receive the training.

Summary

In summary, the first responders working today have serious and dangerous jobs. They are constantly placed in dangerous situations and embrace the challenges of their jobs. They are far more fortunate than the responders from the previous one hundred years. Technology has given the professional responder a great many tools that have enhanced their ability to do their jobs safely and efficiently. Given this information, it would seem that our responders are completely trained and equipped to respond to and mitigate any type of emergency. On a routine and daily basis, all of the response communities do a fantastic job in their own areas. However, should a major incident occur—especially one related to terrorism, responders may be caught off guard. Equipment is needed, but so, too, is training. We need to prepare our responders for the unexpected and prepare them to adapt to new challenges. Drastic situations call for drastic measures, as the saying goes. In the case of first responders, a terror attack is a drastic situation. The drastic measure will be the ability to adapt to the crisis, to cross into the traditional territory of other responders, and to do whatever is required to assist victim. If we expect certain responders to overcome these obstacles, we need to provide them with the necessary training and equipment. In times of emergency, victims and their families do not want to hear excuses, they want to see results.

We were already exposed to a crisis of previously unseen dimensions in 2001. We have had time to lick our wounds, to develop new strategies and technologies, and to expand our training base. We have the benefit of studying our weaknesses prior to 9/11. We have spent the money to develop equipment and tools to meet the challenge and we have trained thousands of responders. New technologies are constantly being developed to make responders better equipped to deal with the next incident, whatever the size. Our level of expertise and technology has never been greater and our resolve has been tested as never before. The problems with our training have been identified and solutions and strategies have been suggested. Given the time that has elapsed, there will be no acceptable excuse if we are unprepared to deal with another attack and if we fail to show the world just how far we have come regarding technology, training, and preparedness.

Chapter 15

What Have We Learned?

First responders are a special breed of people. They have serious and important jobs and careers, and literally save the lives of ordinary citizens on a daily basis. The rewards are mostly internal and even career responders never become wealthy on their public salaries. With each type of response comes varying levels of excitement, danger, liability and challenges. All over the world, people face crises on a daily basis and when they need assistance, the first responder community always answers the call.

There are four major sources of first responders: law enforcement officers, firefighters, emergency medical responders, and hazardous materials teams. Each is master of his or her own domain, but some responders receive training into other response disciplines. The training that crosses disciplines is very basic and each response community faces training and certification requirements. Of all of the responders, the group most at peril is law enforcement. These responders are dispatched to every type of emergency service call, yet they have virtually no training outside their realm. Police are routinely sent to fire calls, emergency medical calls and hazardous materials calls where hazardous materials are present or have been released. These responders have the fewest tactical tools to deal with various emergencies, and are required to request additional, better qualified resources to the scent to mitigate the incident.

Law Enforcement Review

While law enforcement officers will generally arrive first, there are incidents where their mere presence will do nothing to mitigate the incident. Upon arrival, the police officers are able to provide information to other responders that can be passed along to them by dispatchers. This may not seem important to the average citizen, but the information provided to other responders will have a profound effect on the resolution of the emergency. But what happens when law enforcement officers feel compelled, or duty bound to act in areas outside their traditional boundaries? Sometimes nothing bad happens, but sometimes officers are injured or killed due to a lack of training. While not all emergency responder deaths can be avoided by cross-training, there are some that could be, and society owes emergency responder the best equipment and training available while they serve their community. Expectations are high once a call for emergency assistance is made and, regardless of the emergency or the responders who arrive, people demand service. First responders are problem solvers, and our "clients" demand that we resolve their crisis in a timely manner. In order for this to be a realistic expectation, cross-training may be the solution.

In reviewing the need for cross-training, it is obvious that law enforcement personnel are the most in need. They are at all service calls, and are generally first to arrive. A major area of need for the law enforcement community is in regard to fire and hazardous materials responses. All too often, police officers become victims at these types of responses. The reason is quite simple: no meaningful training is provided to these officers. Simple instruction on the science of fire may enable officers to avoid allowing a fire to grow in intensity or to hinder the efforts of fire crews. Taking additional time when arriving at fire incidents to be certain that fire hydrants are not blocked can also assist fire crews. Understanding that police uniforms will melt, and that searching for victims in a *real* fire, are very different than as portrayed on television and in movies will also lead to safer police operations. Aggressive police response to hazardous materials incidents can be disastrous for the officers. They need better training to understand response dynamics, including factoring in wind direction prior to choosing an approach strategy. Sometimes the safest and most prudent action is to do nothing, and this may be difficult for some responders to understand or accept. Law enforcement needs regular training in the areas of fire service activities and hazardous materials response.

Another area where law enforcement is lacking is in terrorism response. The complacency shown since 9/11 is scary. While we will never forget the events of that horrible day, the facts are no longer in the forefront of the minds of our leaders. Other issues, such as gangs, drug sales, murders, and other daily criminal events have forced terrorism response to the rear. This is unacceptable and once our guard is completely lowered, we will be ripe for another attack. We have been afforded a great deal of time to ready ourselves for such an incident and we were very diligent in 2002 and 2003. However, as the calendar has moved forward without further incident, our priorities have shifted. While we are not as reactive as we once were, we are slipping closer toward reacting rather than being proactive.

All first responders need to be exposed to information and training normally reserved for other first responders. The terrorist groups are here to stay and our professional responders need to be ever vigilant and prepared to act on a moment's notice. The broad fields of terrorism response, IED awareness, suicide bomber awareness, and CBRNE need to be expanded and presented to all emergency responders. Under stress, human beings will resort to their training, and experiences. If there are no experiences, the responder is in a bad tactical position. Consider the "training" drivers have in their vehicles. If a child or animal darts into traffic in front of a car, the driver automatically applies the brakes. The driver doesn't need to look down at the pedals or wonder where the brakes are. The body and mind act as one, and reflexively, almost without thinking, the brakes are applied. Regardless of the scenario, responders need this type of reaction in times of crisis. Training levels should prepare all responders to react to any type of emergency that may arise. The best way to reach this level of preparation is to subject responders to training across their own response discipline. Preparation is the key to keeping one's cool under pressure and this level of calm is essential to make and deploy sound decisions and tactics. Responding to an incident without prior exposure results in delayed decision making, and this delay can prove deadly.

But are there others in the broad community of responders who need training as well? There most certainly are, and they have already been identified. There area great many assets in society that can be called upon in times of emergency. Some of the identified resources already consider themselves to be first responders, but are untrained to a great extent. Two of the largest groups of potential responders are security guards and private EMS providers. Security personnel in malls, hospitals, chemical plants, and other critical locations can be of great assistance. Security guards are generally found in soft target locations or locations where there are no target hardening measures in place such as bollards, security checkpoints, etc. Malls, casinos, schools, and banquet facilities are all examples of soft targets. Most of these sites employ private security personnel on-location. If properly trained, these security officers can prove invaluable in gathering information. Additionally, should an incident occur, these on-site responders will arrive even prior to law enforcement. Their observations will provide the initial framework for the response of other professionals. Investing a little time to train these officers in terrorism awareness, IED and suicide bomber awareness can have a tremendous upside, and assist planners and first responders. A good deal of the security guard community is composed of young people looking to become law enforcement professionals. Their enthusiasm should be noted and utilized. Other security guards may be retired law enforcement officers seeking to enhance their pensions. These security officers are already trained and can be assets at incidents.

Private Resources

Private companies dealing in medical transports are another untapped, untrained pool of potential first responders. At mass casualty incidents, there will be a tremendous need for EMTs and ambulances for transporting the victims. In the incident management environment, it is recognized that there will not be enough personnel or vehicles for patient transport. A good method of securing these resources would be to contact these private medical transport companies. The employees transporting the patients on a daily basis are most likely EMTs. These trained medical responders can be recruited to assist in many instances. While they are private company employees, they are also trained and available to serve as needed, if the employer will release them to the incident. The ability of incident managers to recognize that these responders are available will have a direct impact on the resolution of the incident.

Providing these employees some additional training at their orientation can be a worthwhile investment for incident managers. These trained medical responders can be requested to the scene and will enhance the treatment and transport capabilities of incident managers. Providing them with additional training on signs and symptoms and treatment options for chemical exposures and basic terrorism awareness along with suicide bomber awareness is all that would be required.

Airport Resources

In addition to the traditional responders, other potential first responders are working every day in a variety of fields. Airport security personnel such as Transportation Security Administration (TSA) security screeners are an example. These professionals are trained in areas including security screening techniques, metal detector machines, x-ray equipment, and agency policy. The use of commercial aircraft may still be seen as targets of opportunities for certain terrorist groups. With this in mind, constant training of these security personnel in areas including body language, potential indicators of violent or criminal activities, and responding to incidents would be desirable.

In years past, the airport terminals themselves were areas where violence and terrorist activities were conducted. Should this trend re-emerge, the TSA security staff will be likely called upon to assist the on-site law enforcement officers in responding to the incident. In the confusion that would most likely occur during and immediately after an act of terror in an airport terminal, swift and decisive action will be required to restore order and maintain strict security controls. For example, there have been many incidents inside airport terminals. One such example was the July 4, 2002 shooting inside Los Angeles International Airport at the El Al ticket counter. During this incident, a forty-one year-old Egyptian who had been living in the United States for two years killed three people and injured four others.[1] In this incident, there was no further plot by any other individuals or groups, but consider if the incident had been part of a larger plot. What if the shooting had taken place closer to the security screeners? Could a random shooting be a "distraction" to get TSA personnel and law enforcement personnel away from their posts, thus allowing co-conspirators access to the secure area during the confusion? This scenario is certainly possible and the likelihood of success would depend on the ability of TSA and law enforcement personnel quickly mitigating the distraction, while simultaneously preventing a breach in security. In this case, TSA personnel would serve as possible first responders and as terrorism preventers by not allowing the "distraction" to result in the security breach. These professionals are the last line of defense for the civilian air transportation industry and, as such, need constant training. These security screeners will continue to have an important role in trying to prevent terrorism in airports and, more importantly, on aircraft once they leave the protective environments of the terminals.

School Resources

Children and parents expect that they will be safe while attending school. Although this would seem to be a reasonable expectation, it is clear that schools have been targeted not only by terrorist groups, but lone or single issue terrorists as well. Parents entrust the safety of their children to school teachers and administrators. In the event that any school or student comes under attack, the first responding adults will most likely include teachers and administrators. The teachers in schools in recent years have been given specific instructions regarding procedures to be followed in the event of a school shooting and sadly, there have been a number of school shootings where these procedures have been tested. Teachers need to be trained and capable of handling these types of incidents to ensure the safety of the children in their care. Administrators need to be trained to be aware of characteristics and indicators that an incident in their schools might be imminent.

There are classes available that deal specifically with these issues and law enforcement has offered training to both school officials, and the law enforcement community. Police and SWAT teams are now trained to respond to active shooting incidents in schools. The school administrators also need to be trained to "harden" the security at the schools and even engage local law enforcement to provide a *school resource officer*. It is sad that schools need to consider this option and sadder yet that we need to make first responders out of teachers, aides, and administrators. However, the problem has been clearly identified, and to turn a blind eye to the potential that this type of incident can happen at any school at any time is unacceptable. Parents and children demand safety while the schools are operating and we need to assist teachers and administrators in accomplishing this mission. The employees of our school system need to understand that they may be called upon to be the first responder at any time and they need to be trained accordingly.

Public Health Resources

Public health officials may also be called upon to assist the first response community. After the September 2001 attacks, biological incidents and threats became common. In the United States, there were biological incidents involving anthrax in Florida and New Jersey. These incidents resulted in one death, and in New Jersey, a post office facility was closed for several years, requiring a thorough decontamination of the entire facility prior to reopening for business. The anthrax used in the incidents was mailed to intended targets. The response community was besieged with what were loosely termed *white powder jobs.* These incidents referred to any call for service where the caller reported a white powdery substance. Obviously, almost all of these incidents involved substances other than anthrax. The public health community was very active and involved in these responses, and their roles have been discussed. They are very helpful in determining if the incidents involving public health are focused responses, or if they are considered public health emergencies. Providing public health officials additional training across traditional boundaries should be carefully considered and implemented.

OEM and Its Roles

Once an event or incident occurs, the response community is activated and begins the process of mitigating the incident. If an incident grows too large for the initial responders to handle, additional resources are summoned to the scene. For very involved and complex incidents, the Office of Emergency Management will respond and assist. These people generally serve in the capacity of logistics, and assist incident managers in ordering, tracking, and releasing resources. The process of orderly processing requests for certain assets to an emergency scene is critical to the safe and efficient operations of those managing the incident.

OEM personnel need to train with all branches of emergency service, in order to be in the best position to provide timely assistance during the incident. If these personnel have been cross-trained in CBRNE, hazardous materials response, and terrorism awareness, they will be able to anticipate the needs of managers during the incident. They also need to train with fire and other responders to understand the capabilities and limitations of the responders. By being exposed to training scenarios and observing what resources will be required to combat each type of emergency, they will not be caught off guard in times of real emergency. OEM staff may then be in a position to have equipment and vendor lists readily available for use during incidents, thus reducing the time required to locate and order certain specialized equipment. These specialized resources are referred to as *critical resources* under the incident management system. OEM officials will need to have the ability to locate and order multiple critical resources should a large-scale incident occur. Examples of critical resources include Medivac helicopters, hazardous materials teams, and bomb squads. Having trained with all response organizations, and being exposed to all available training classes, the OEM staff will be put in the best possible position to provide responders with all necessary equipment.

Politicians and Their Roles

First responders are always ready and willing to answer the many calls for service presented to them. These responders spend tremendous amounts of time away from family members and friends, especially those in the volunteer ranks. Regardless of whether the responder is a career member or a volunteer, he or she will require minimum training and equipment. Inevitably, the political leaders must become involved in this process. Politicians, especially at the local level, always manage to respond to local emergencies, and are usually interviewed by the media for comment on the incident. As with most leaders in society, if a good job is done, or lives are saved, the politicians are happy to be interviewed. However, if the incident goes badly, or there is loss of life, the media finds interviews harder to come by.

There is an expectation by our political leaders that any incident can be managed and controlled by the response communities within their jurisdiction. Sometimes, these expectations are unreasonable and, unless these leaders are exposed to our training, they simply will not understand our limitations. Involving politicians in the training process can have many positive results, and can lead to a better working relationship between responders and government. When our local leaders see the equipment that we have, and then are exposed to the equipment that will be needed, they may be in a better position to offer support. Inviting them to observe when training is undertaken may also elicit empathy from these leaders, and they may develop a greater appreciation and understanding of exactly what first responders actually do.

As previously noted, the entertainment industry has led many people to expect that crimes can be quickly solved, that most medical patients can be resuscitated, and that fires are not unsafe environments. Having the politicians experience the real-world emergencies is bound to open their eyes. Should they see that the needed equipment for managing incidents is unavailable, they will be more willing to make the necessary monetary commitment. Allowing or even requesting that they sit in on classroom training in certain response areas will also help them develop a better understanding of the challenges faced by the response communities. It is important to expose the leaders to the world to emergency response and mitigation training. Without experiencing it first hand, fallacies and misunderstandings will continue to go unchecked. Garnering support from our leaders, especially with regards to budgetary issues will certainly be beneficial to the incident responders and managers.

Training Mandates: Making Cross-Training a Priority

Cross-training can be accomplished in a variety of different ways. There are tremendous numbers of responders who are already cross-trained, and these people, if so qualified, can serve as instructors or trainers. If there are no instructors from your agency available locally, you might ask sister organizations if they have anyone who might be willing to assist your agency. If there are no real solutions for your agency, the next option for classroom training is to send your members to dedicated training academies. Courses may already exist or they can be developed by qualified instructors from other response disciplines. These instructors will need to know the content required as well as the goals, and the time allotted. Most likely, the desired training program already exists in some form and modifications may be required to meet your specific needs.

Training Resources Available to Responders

Other resources are available to assist responders to receive training. There are many organizations and entities that can provide on-site training for your organization, at no cost. These training groups were previously discussed in Chapter 10. Most of these training agencies are funded from federal coffers and specific classes have already been developed and are ready for presentation. Agencies such as Texas Engineering and Extension (TEEX) will offer incident management classes for example. Other organizations will come to your agency to train your staff in radiological response, biological and chemical response, and provide classes related to suicide bombers and IED. Given that these organizations are fully funded, have developed training curricula, and will bring their classes to you, there is no excuse for your staff to not be trained. Additionally, there is the ability for agencies to allow certain members to report to specialized training sites across America, and enjoy quality training and education at no cost to the employee or agency.

Very few agencies are aware of these training opportunities, and investigation into them can enable the agency to send an employee or two to the class, and bring the class back to the agency as a trainer. Failure to explore and avail your staff of these training options is completely unacceptable, given the cost. Reporting to these classes allows your members to be exposed not only to classroom education, but follow up hands-on training at the facility.

Funding Training

While the federal government has generously provided certain organizations and facilities with the funding to provide free training, not all required or desired training is free. In the current financial climate, administrators battle with budget shortfalls, manpower shortages, and uncontrolled overtime. Requesting additional time for training, especially cross-training, may be met with resistance by administrators. Should this happen, the need for the training must be explained. For example, a current phrase uttered all too often is that we are doing more with less. Use this thought process to drive home your need for training. Fewer responders will mean those remaining will have to do more to make up for their reduced numbers. It has already been shown that all responders are crossing their traditional training boundaries in an effort to cover for depleted resources in the field. If your agency requires you to provide service outside your area of traditional and practical expertise, you must demand that you receive adequate training in the new area of responsibility. Liability issues alone will dictate that responders have training prior to providing service in an emergency situation.

Responders in the field still respond to emergency calls, and regardless of the state of their agency's finances, they provide the best possible service. If administrators want those remaining resources to do more with less, they must allow the training process to be followed. If this results in overtime or increased training budgets, this is the price for allowing staffing levels to drop below certain minimum allowable standards for your agency. Administrators deal with politicians and the budgetary processes, and the employees, volunteer or career, are charged with mitigating the emergency calls for service. Everyone in the system must focus on his or her individual responsibilities, and handle each call for service in the most professional and practically efficient manner.

Online and Distance Learning Review

The form of training staff in all response disciplines varies. The last area of training can be conducted in-house, and will not require any cost or instructors. This method will involve the use of computer-driven or virtual academies. This venue for training has its share of advantages and disadvantages. The advantages are apparent and many. The classes are taken online, and the classroom never closes. Therefore, responders working various shifts can study and "attend" class during their regular tours of duty eliminating the need to rotate personnel to different shifts, and avoiding overtime costs. Since there are no live trainers, classes can be taken at the pace most suitable for each student. Courses may be taken over several days and this ability to stop and restart allows for flexibility by management. Another advantage of online courses is that different sites offer the courses. FEMA, the National Fire Academy, and most states have online course offerings. These classes provide a good core of material for the willing student. Some states and agencies have mandated certain training classes be taken via these virtual academies.

While these sites offer a wide array of training and provide great flexibility, there are also drawbacks. Certain types of training, such as incident command classes rely very heavily on practical scenarios to test the student's understanding of the material presented. In an instructor led environment, these scenarios are easily accomplished by separating the class into several working groups. The scenario is presented, and groups are provided ample time to review and develop appropriate responses to the scenario. Once the groups are finished, the instructor reviews the proposed solutions offered by each group. This review process reinforces the concept that there are different possible solutions to the issues raised in each scenario, and that no one answer is completely correct or completely incorrect.

In a forum where the class is taken online, there is no way to simulate these scenarios, and evaluate the proposed solutions. The students are deprived of seeing the different strategies and structures that may be chosen for an incident or event. The scenarios are the best manner for instructors to determine if the students are grasping the material, and also to ensure that the materials can be properly applied to an incident. Additionally, instructors in classroom settings are able to answer any questions that their students may have. The virtual academy style of learning offers no mechanism for questions and answers. This is a very negative aspect of this style of instruction. Even if there were a forum for students to ask questions, the correct manner to answer the question would be immediately and allow for a follow-up question if doubt remains. Receiving an answer to a question after several hours or days deprives the student of the ability for immediate response to ensure complete understanding. This drawback needs to be considered for all types of Internet based classroom learning. If students truly desire to learn subject material, there exists a great likelihood of questions being posed.

The final drawback involves classes where tests are required to receive credit for the class. While no one wants to admit it, most agencies that require virtual learning have test answers available for their staff. There will always be people looking for a shortcut or a way out of undesirable assignments. If the required training is of no interest to a student, chances are that nothing will be gained and those desirous of avoiding the training will solicit test answers from their peers or colleagues. Administrators must be aware of this possibility, and take steps to prevent it from happening. People who log on to the site, and leave the building only to return later to take an exam get no benefit from the training class. In fact, it will have the opposite effect. If administrators or supervisory personnel believe that a member of their staff has a competency that they do not, there will be issues later. Also, people are far more likely to forget the materials presented in this forum as opposed to an instructor-led environment. Accordingly, virtual academies and online mandated course offerings need to be closely monitored.

Many agencies have staff members who wish to continue their formal education. With work hours, rotating schedules, overtime, and family issues, attending an institute of higher learning is not always practical. Recognizing this dilemma, many universities and colleges have turned to a concept known as distance learning. This entails a student taking college classes from home and completing the bulk of the course work online. This is completely different from the virtual academies. These courses are real college classes, and the work is submitted to live teachers for grading and evaluation. The ability to take college classes online is very desirable for the emergency responder of today. Many employers place a premium on formal education, especially for promotional opportunities. These educational opportunities are commendable and members seeking to better educate themselves should be encouraged. In addition, the cost for these courses is much less than actually reporting to a classroom for formal instructor-led classes. Students should be cautioned to investigate their college of choice to ensure that it is properly accredited. Since this is formal education rather than mandatory brief classes, the distance learning educational process is encouraged, and the issues associated with virtual academies are moot. Students will have contact with their professors, and have a forum wherein they will be able to ask questions and get answers.

Grant Writing Review

After an agency is able to provide training for their members, there may be a need for new equipment and materials in order to meet this new mission. Often, agencies branching into new training areas are not in a position to equip their members with all of the required material as a result of financial restraints. In these cases, the agency should solicit grant funding. Many different agencies and programs offer funding opportunities for all branches of emergency response. Grant opportunities can be researched online via any reputable search engine. When applying for these grants, certain fine points need to be closely observed. Applicants must be certain that they are eligible to apply for the grant sought. For example, some grants are limited to small agencies, some only offer certain equipment or materials, and some may be regional. Another consideration is to be certain that the grant application period is active. Each grant has an opening and closing date, and these dates are strictly enforced. Also, the person who completes the grant application must pay close attention to all of the instructions. Failure to comply with the instructions generally results in a quick rejection. Certain grants offered to organizations will also require a financial match. This means that the awarding authority will pay a portion of the total approved grant award, and the remaining balance must be paid for by the organization awarded the grant. This usually requires that the applicant acknowledge that a match is required, and that the matching funds will be available should the grant be awarded. If there is not sufficient money in your budget for this match, it is imperative that your governing authority agree to cover it. Failure to pay the matching portion of the grant will result in the award being pulled from the receiving agency. If a narrative is required, as they almost always are, this is the area where your application will be most closely reviewed. If you are seeking funding for a new operational mission, you must make this abundantly clear in your narrative, and detail why you are expanding into this area, as well as the financial restrictions placed on your agency.

Another area where great consideration is afforded is where safety or standard compliance is a stated goal. For example, if a fire department needs to replace an aging fleet of SCBA, the narrative should cite the applicable NFPA standard. Further, the narrative should indicate the goal of achieving one hundred percent compliance with the standard, and include the current percent of compliance. While it is not necessary to retain a professional grant writer, whoever completes the application needs to be certain that it is accurately and completely filled out. Sometimes it is better to have an employee or staff member author the grant, as they are intimately familiar with agency's parameters, finances, and operational shortcomings. Grant opportunities are the most financially beneficial ways to ensure that your agency has all of the required tools and equipment needed to complete their mission.

Lastly, grant opportunities also exist for training sessions as well, and if your agency cannot afford to offer certain classes, the grant will allow the agency to provide the required training. These opportunities are separate and apart from the training classes that were previously mentioned, and the off site classes as well.

Testing What We Have Learned

To this point, a review has been offered to cover who needs to be trained and why, what areas of training need to be offered, how to get and pay for training, and grant opportunities for your agency. Once the agency has received the training and any necessary equipment, the next logical step will be to test the training. This can be accomplished in either of two manners. The first was to design and execute a table top drill scenario. This process is a very long, tedious engagement, but there are benefits to the process. The primary benefit is that participating agencies can model, or simulate certain tasks, without the need to actually have the positions staffed. This can reduce the size of the drill, and save money since on duty personnel or resources are not required to be present for the drill. The drawback is that these responders are not present. We will address the hands-on drill shortly.

There are many steps that need to be taken prior to the actual drill going live. A goal for the exercise needs to be developed and tracked. Participating agencies need to be identified and they need to commit to the entire process. Next, the planners need to decide on a scenario for the exercise. Once the scenario is chosen, it needs to be developed, and moderators need to be selected. The moderators interject factual occurrences into the scenario, as the drill is underway. This allows the planners and managers of the drill to be tested without prior knowledge of changes in the scenario. Observers need to be chosen. Observers must have experience in the field on which the scenario is based. They will simply watch as the drill events occur and take notes for review after the drill is completed. Next, an incident management, or incident command structure needs to be developed. At this point, all critical overhead, or supervisory positions must be filled. This is the point in the process where modeling or simulating less critical positions can be accomplished. The scenario timeline is then developed, and released to participants. This scenario guideline will be intentionally vague to allow for the incident managers to be tested as to their decision making prowess. A date and time will be finalized, and the drill can then be executed. The moderators infuse certain material and issues into the drill, observers watch and participants address all issues raised. Once the drill is finished, a debriefing of on-hand personnel will be conducted, and comments memorialized. A formal review will be scheduled and conducted. All personnel attending the review must check their egos at the door, and realize that the review is not a personal affront or attack. The purpose of the review is to address the shortcomings identified in the drill and the existing guidelines to manage that type of incident. All participants able to attend should be present and all questions should be addressed. Next, any emergency annexes or policies that need to be amended or updated need to be corrected. Once all of that is accomplished, the next tabletop or hands on drill can be scheduled.

Reinforcing the Training

Often, once the table top scenario has been completed, a hands-on drill is scheduled. This will generally not be one in which the same incident as was tested via tabletop is chosen. The initial and follow up steps for the hands-on drill are virtually the same sequence as with the tabletop. The primary differences lie in the areas of manpower. In the formal, hands-on drill, there is much less simulation. The goal here is to test all responders and to determine if the existing emergency plans and standard operation guidelines of the participating agencies are adequate. These drills need not be conducted in real time and more observers may be needed. The additional information infused into the drill by the moderators can trickle down to the responders in the field, and their decisions and tactical priorities can be reviewed. Once the drill is completed, the same review process will be followed. There will most likely be far more questions generated at this review because there are more participants. Review moderators need to keep the review moving forward, while allowing the free exchange of ideas and the presenting and answering of questions. As with the table top review, any annexes tested, or policies that are deemed deficient need to be addressed. The adjusting, amending or creation of new policies and procedures will need to be addressed. All agency personnel will need to receive a written copy of the new procedures. Often, the creation of new policies or procedures requires additional training for affected personnel. The new policies cannot be fully implemented until all affected staff has received the requisite training. At this point, the next exercise can be planned.

Cross-Training Review

Throughout this text, a position was proffered that the first responders of the world need to be cross-trained. Examples were presented to demonstrate where each branch of responder could be better trained, and how these new training areas would benefit society. Only the terrorist or terror group will know when and where the next attack will happen. Our responders are willing and ready to respond to any emergency call for service at any time. Some will respond as paid career responders, and others as volunteers. The victims in need of assistance will not care if their rescuers are one or the other.

Every response branch in the emergency service community is at a crossroad. Budgets are shrinking, manpower numbers are declining and more and more is expected of those who remain. Volunteerism is on a downward spiral for many reasons, including the financial needs of families, training and time requirements, time away from loved ones and friends. There is a trend currently in which formerly all volunteer agencies have been forced to convert to paid or partially paid departments in order to provide the required service to their community. This conversion will have trickle down effects. Taxes will increase, resulting in residents requiring more money to meet this increase. A second job may be required, which will take away any free time, and reduce the slight potential for volunteers to join response agencies. With declining manpower available and governing bodies demonstrating reluctance to hire new personnel, those still on the job are facing daily issues in terms of their ability to do their jobs. Staffing issues also have their trickle down effects on other responders. Fewer EMT volunteers will necessitate the police officer being on scene treating patients longer than before. Smaller fire departments may have slower in-service and arrival times, also requiring law enforcement to remain on scene, and operate longer without the proper resources.

What does all of this have to do with cross-training? Simply put, the resource that arrives first may not be the one requested by the victim. The first arriving unit, from whatever branch, will be asked to address the emergency. Since that unit will be likely waiting for the correct resource to arrive, its members should receive some training to allow them to begin initial resolution of the emergency. Remember, taxpayers and administrators are demanding that emergency services do more with less. While responders may be forced to work outside their occupational and traditional boundaries, they should have the support of the citizens and administrators. A person experiencing a heart attack would not go to the office of a psychiatrist, or an auto mechanic. They would certainly desire to be seen by an emergency room physician or a cardiologist. Why then do citizens expect police officers to enter burning buildings, or respond to hazardous materials incidents? They expect that the police officer is an *emergency service responder.* That fact alone generates an expectation that service can be provided by that responder. A sense of duty will sometimes arise, and improperly trained responders will attempt to *help* the person in need. This unfortunately sometimes can have very tragic consequences for the rescuer and the victim.

Why would a police officer who cannot swim jump into a swollen creek to try to save a child in the water? He has a sense of duty and responsibility, and possibly a feeling that he will overcome his shortcomings. If we are going to have certain expectations of our responders, we need them to be realistic.

Given training, the insistence that any responding agency assist us is not totally unrealistic. However, this is not yet the case. Investing some time and money to put responders in the best possible position to render proper aid is certainly a small price for a society to pay. If another terrorist incident of any significant magnitude occurs, you can be certain that the public will demand a better response than we may be ready to provide. We cannot expect untrained resources to make immediate, life altering decisions, or to develop strategies to deal with incidents without a scintilla of related training.

While the expectation may be unrealistic as viewed by the responders, it will be very realistic to the public. They have watched the government spend exorbitant amounts of money on homeland security. They have watched tax dollars flow into the hands of emergency planners and responders, enabling them to purchase new tools and equipment. Response issues have been identified, such as a need for interoperable communications. We have been afforded a window of more than eight years to address these issues and prepare ourselves for any attack. Eight years. A demanding public would expect a better response after eight weeks, let alone eight years.

But are we truly ready to respond to an attack with the proficiency that society expects? Our human resources are dwindling, and their duties are expanding. Overtime costs and mandated training for each branch of emergency service make cross-training difficult. Society will not want excuses such as these if terror strikes. They will demand that proper, immediate actions be taken.

Identifying the Problem and Offering Solutions

The issues regarding cross-training have been identified in this text. The solutions, while not easily and readily implemented, have also been identified. Ignorance of the problem can no longer be tolerated. The response community has been dealing with these issues for years, and no significant strides have been made. The priority of cross-training grows less and less with each passing day. Each response branch has its own issues to deal with, and expanding roles and training are not seen as important issues. Complacency is a formidable foe, and society as a whole is slipping into that mode. We cannot forget that there was an eight-year gap between the World Trade Center bombing in 1993 and the 2001 simultaneous attacks on America. The terrorist is a very patient foe, and our society has no patience. We have been lulled into a false sense of security by a variety of factors. We have witnessed the *war on terrorism* as seen on television almost nightly. Congress has appropriated billions of dollars to prepare America for the next attack, and we believe that we have somehow hardened America. The United States has developed Urban Area Security Initiatives (UASI) in some states. These areas have been given additional homeland security funds to address the potential for terrorism.

Given that these urban areas are identified and commented on by the media, does anyone think our enemies are unaware of these areas?

Is it possible they will target areas far away from these secure areas, and look for untraditional target sites?

That possibility is quite real. Residents in these rural areas, considered "in the middle of nowhere," probably feel quite safe, since they are far away from large metropolitan centers. But are they?

Have we neglected these small out of the way places to the exclusion of the large cities?

Can these small towns and hamlets absorb the effect of a terrorist attack?

Remaining Vigilant, Studying Trends, Identifying Deficiencies

Current trends in terrorism indicate that when an attack is planned, there is more than one target site. This was evident in the London subway attacks, the Jordan hotel attacks, and in America, on September 11. We must realize that this was a coordinated, simultaneous attack. Secondary attack ideologies include the killing of emergency responders. In New York City, that goal was certainly met. The city of New York lost more than 300 responders in one day simply because they had that many resources available who were working to save lives. Can a rural area in Kansas handle a simultaneous attack scenario? What about a large city such as Topeka, or Kansas City? We need to train our first responders across the country in as many areas as we can. This training must be considered a priority, and we must begin to accomplish this aggressive program immediately. We have had a eight-year head start, and society will not understand if the emergency response system comes up short again.

There are "never forget" signs and stickers all around us. While we can easily remember what occurred if prompted, it is no longer in the forefront of our minds. This is what the enemy wants and that is what they have gotten. While it is understood and agreed that each jurisdiction faces other problems and issues, it is unacceptable to pretend that we are prepared and hoping there is no need to prove it.

While I certainly do not know when and where we will be subject to an attack, I do know that we need to be better trained and prepared. Classroom training alone will not be enough during a response of great magnitude. Responders need to be exposed to the hands-on drills to reinforce the principles and techniques discussed in the academic environment of the classroom. There is no substitute for experience, especially practical experience. Since a response to an act of terrorism is a rare task, responders need to have had some practical exposure to ensure that they effectively handle the duties they will be performing. Our emergency service community will answer the call when we are attacked. The issues facing the responders are apparent, and have been identified. I just hope and pray that those responders will be adequately trained and equipped to deal with what awaits them. Solutions to these issues have been presented, and it is incumbent upon our administrators and government leaders to act on them. Failure is not acceptable.

Endnotes

1. "Los Angeles airport shooting kills 3" Article published July 5, 2002. Available at: http://archives.cnn.com/2002/US/07/04/la.airport.shooting/

Index